California Science

Science Content Support
Grade 3

Harcourt
SCHOOL PUBLISHERS

Visit *The Learning Site!*
www.harcourtschool.com

Printed in the United States of America

ISBN-13: 978-0-15-352282-6
ISBN-10: 0-15-352282-8

2 3 4 5 6 7 8 9 10 022 17 16 15 14 13 12 11 10 09 08

Contents

Getting Ready for Science

Unit 1 • Energy and Matter

Unit 2 • Light

Unit 4 • Patterns in the Sky

Vocabulary

Name _____

Date _____

Welcome to *Science* by Harcourt School Publishers. You can look forward to an exciting year of discovery.

Your textbook has many features that can help you learn science this year. Use this scavenger hunt to learn more about it.

1. What animal is on the cover of your book? _____

   Name one fact about the animal. _____

2. Find the unit called "Getting Ready for Science." Name one of the science tools you will use this year. _____

3. What are the three handbooks in the back of your book?

   _____

4. The Big Idea is what you will understand by the end of the unit. What's the Big Idea for Unit 4? _____

   _____

5. In order to understand the Big Idea, you will be answering Essential Questions. The Essential Questions are also the titles of the lessons. What is one of the Essential Questions for Unit 3?

   _____

6. What Reading Focus Skill is used in Lesson 4 of Unit 3? _____

7. What is the first word in the glossary? _____

8. Find an activity that you would like to try on one of the Make Connections pages. Write its title and page number.

   _____

© Harcourt

9. Name the title and page number of an Insta-Lab you would like to try.

_____

10. Find a California on Location feature that you find interesting. What's the location? _____

11. Find the postcard at the beginning of Unit 3. Who is it from? Where was it sent from? _____

12. You can use the Vocabulary Preview to learn new science words from each lesson. The preview shows you how to say, or pronounce, the term. Find the Vocabulary Preview for Unit 2, Lesson 1. List two vocabulary words that you will learn. What photos are shown to help you understand those words? _____

_____

13. The words in the Vocabulary Preview also appear in the text of your book. They are highlighted in yellow and are used in a way that helps to explain their meanings. Find the two words from the item above in the text and list the page number each appears on. _____

14. Find a California Fast Fact that you find interesting. Write its title and page number. _____

15. Write the name of a person featured in one of the People in Science features. _____

16. Write three new things you expect to learn about this year.

_____

_____

_____

Name _____

Date _____

Lesson 1—What Are Some Science Inquiry Tools?

Vocabulary Power: Context Clues

Read each sentence. Think about the meaning of the underlined word. Write the meaning on the line.

1. In his <u>inquiry</u>, Jorge used a hand lens to find out more about plants.

_____

2. Chantel carefully squeezed the <u>forceps</u> to lift the butterfly wing she discovered on the ground.

_____

3. Brian measured the length of each beetle three times to be certain that his measurements were <u>accurate</u>.

_____

4. Josephina read the <u>thermometer</u> to find out how cold the water was.

_____

5. Sara could see each hair on the beetle when she looked at it using a <u>hand lens</u>.

_____

© Harcourt

Use with Getting Ready for Science.   **Science Content Support** CS 1

Name _____

Date _____

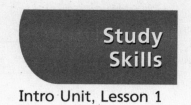

Lesson 1—What Are Some Science Inquiry Tools?

Take Notes

Taking notes can help you remember important ideas.

- Write only important facts and ideas. Use your own words. You do not have to write in complete sentences.
- One way to organize notes is in a chart. Write down the main ideas in one column and facts in another.

As you read this lesson, use the two-column chart below to take notes.

What Are Some Science Inquiry Tools?	
Main Ideas	**Facts**
Tools Used for Inquiry • Tools used to collect data • Help you to observe, measure, and compare objects	• Hand lenses make objects look larger • _____ • _____ • _____
Measuring Tools • _____ _____ • _____ _____	• _____ _____ • _____ _____
Some Other Tools Used in Science • _____ _____ • _____ _____	• _____ _____ • _____ _____

© Harcourt

Name _____

Date _____

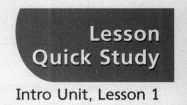
Lesson 1—What Are Some Science Inquiry Tools?

1. Investigation Skill Practice–Observe

Mia wanted to observe a strand of hair. Which tools from the box below could she select to make her observation? Why should she choose these tools?

thermometer	ruler	microscope	measuring cup
spring scale	hand lens	dropper	graduate
forceps	measuring tape		

_____

_____

_____

2. Focus Skill   Reading Skill Practice–Main Idea and Details

Read the selection. Underline the main idea. List at least two details about the main idea.

It is important to be careful using science tools. Many of the tools cost a lot of money. The tools may break if you do not use them properly. You could get hurt if you don't follow the safety rules for using the tools.

_____

_____

© Harcourt

Name _____

3. In the chart below, write an object in the second column to tell what the science inquiry tool can be used to observe.

Tool	What I Can Observe With It
hand lens	spots on a ladybug
dropper	
forceps	
magnifying box	
spring scale	
thermometer	
ruler	
measuring tape	
microscope	

4. Janice has a jug of orange juice that she wants to share with 3 friends. What tool can she use to accurately divide the juice into equal portions? Explain how she would use her tool.

_____

_____

_____

_____

Name _____

Date _____

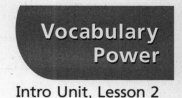

Lesson 2—What Are Some Science Inquiry Skills?

Written Cloze Exercise

Study the words and their definitions below. Then fill in the blanks with one of the words. Use all of the words once.

predict: to tell what you think will happen in the future

evidence: information collected in a scientific inquiry

opinion: a personal belief that is not based on evidence

model: a copy of something very large or small that scientists make

compare: to identify how things or events are alike and different

observe: to study something closely

1. Miguel wanted to _____ the characteristics of two different hibernating animals.

2. Lucy used a telescope to closely _____ the stars in the night sky.

3. Douglas decided to _____ that the weather would be rainy next Tuesday.

4. Andrew's _____ was that the chocolate tasted too sweet.

5. According to the _____ collected, the biggest rabbit weighed 15 pounds.

6. Jasmine used a _____ to show how planets travel around the sun in space.

© Harcourt

Use with Getting Ready for Science.   **Science Content Support** CS 5

Name _____

Date _____

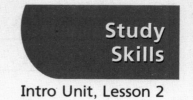
Lesson 2—What Are Some Science Inquiry Skills?

Preview and Question

Identifying main ideas and asking questions about them can help you find important information.

- To preview a lesson, read the titles of each section. Look at the pictures, and read their captions. Try to get an idea of the main topics, and think of questions you have about each topic.
- Read to find the answers to your questions. Then recite, or say, the answers aloud. Finally, review what you have read.

As you read this lesson, fill in the chart and practice reading, reciting, and reviewing.

What Are Some Science Inquiry Skills?				
Preview	**Questions**	**Read**	**Recite**	**Review**
Inquiry skills are used in investigations.	What inquiry skills have I already used in an investigation?	✓	✓	✓

© Harcourt

Name _____

Date _____

Lesson 2—What Are Some Science Inquiry Skills?

1. **Investigation Skill Practice—Predict**

Suppose you had two African violet plants. You kept one plant on a sunny windowsill and the second plant in a dark closet for one month. Predict what would happen to the two plants.

_____

_____

2. (Focus Skill) **Reading Skill Practice—Main Idea and Details**

Read the selection. Underline the main idea. List at least two details about the main idea.

Sharon used modeling clay to make a model of the dinosaur *Stegosaurus*. Models are used in science to study animals that are extinct, such as dinosaurs. The model showed that the *Stegosaurus* had large bony plates. The plates stood up to help the dinosaur control its body temperature. The model also showed that it had long pointed tail spikes. It used these to defend itself against enemies.

_____

_____

_____

_____

© Harcourt

Name _____

3. Fill in the chart below. In the second column, write how
   each science inquiry skill might be used.

Inquiry Skill	How It Might Be Used
observe	the sound a bird makes
order	
use numbers	
measure	
predict	
display data	
compare	
classify	
use models	

4. Two third-grade students observed a skunk and recorded
   their observations in a notebook. One student wrote
   statements based on evidence. The other student wrote
   statements based on opinion. What statement might
   each of the students have written in their notebooks?

   _____

   _____

© Harcourt

Name _____

Date _____

Lesson 3—How Do We Use Numbers in Science?

Graphic Organizer

Fill in the blank sections of the graphic organizer below.

Using Numbers to Display Data		
Ways to Display Data	Definition	An Example of Data It Might Display
data table		how high a ball bounces each time you drop it
bar graph	a graph that uses bars to display data	
	a graph that uses a line to display data	changes in your height over five years

Related Words

Use the words in the box to label the parts of the bar graph.

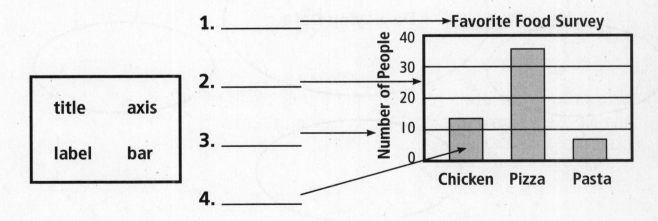

1. _____

2. _____

| title | axis |
| label | bar |

3. _____

4. _____

© Harcourt

Name _____

Date _____

Lesson 3—How Do We Use Numbers in Science?

Connect Ideas

You can use a web organizer to show how different ideas and information are related.

- List important themes in the ovals in the web's center.
- Add ovals showing main ideas that support each theme.
- Add bubbles for the details that support each idea.

Copy and complete the web below as you read this chapter. Fill in each bubble by adding facts and details.

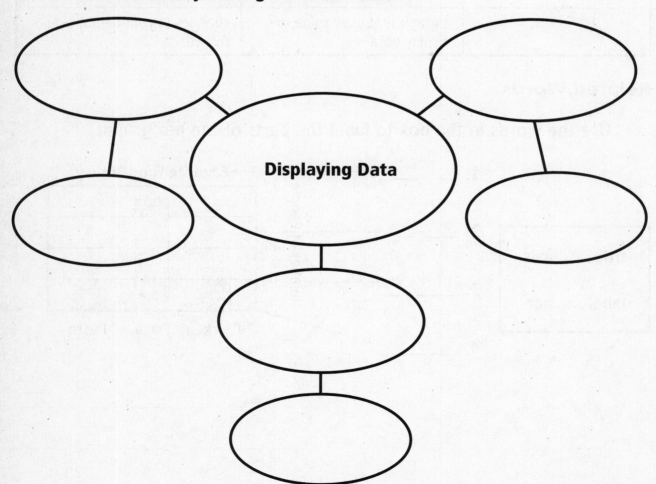

Displaying Data

© Harcourt

Name _____

Date _____

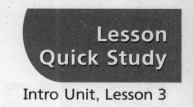
Lesson 3—How Do We Use Numbers in Science?

1. **Investigation Skill Practice—Analyze Data**

Rob gathered the data below so he could make a bar graph.

Name	Weekly Allowance
Rob	$2.95
Tanya	$5.05
Rita	$4.20
Carlos	$3.75

When Rob makes his bar graph, who will have the tallest bar? Who will have the shortest bar? How can you tell?

_____

_____

2. **Focus Skill** **Reading Skill Practice—Main Idea and Details**

Read the selection. Underline the main idea. List at least three details about the main idea.

   We use numbers every day. Numbers tell us how old we are, how tall we are, and how much we weigh. We need numbers to buy items and get back the correct change. Numbers tell us the temperature, the date, and the distances we need to go. Without numbers, we won't be able to learn much!

_____

_____

_____

© Harcourt

Name _____

3. **Check (✓) the statements below that agree with the information found in the lesson.**

    _____ Scientists use numbers to compare things.

    _____ We do not need numbers to measure the distance to the moon.

    _____ You can use minutes to measure the passage of time.

    _____ You can use a ruler to measure how tall you are.

    _____ We use centimeters to measure temperature.

    _____ Each column in a data table needs a heading.

    _____ Bar graphs work for data that is in categories.

    _____ Line graphs cannot show how two different things are related.

    _____ Line graphs use lines and bar graphs use bars to display data.

    _____ Graphs do not need titles or labels.

4. **Why is it important for people to agree that one day is divided into 24 hours, one hour into 60 minutes, and one minute into 60 seconds?**

    _____

    _____

    _____

© Harcourt

Name _____

Date _____

Lesson 4—What Is the Scientific Method?

Matching Exercise

Match the clue on the left to the term on the right.

_____ an organized method that scientists use to conduct a study

_____ a scientific study

_____ a test done to find out if a hypothesis is correct

_____ a possible answer to a question

_____ a decision based on what you know and on your results

_____ to use your senses to gather information

A. experiment

B. conclusion

C. hypothesis

D. investigation

E. scientific method

F. observe

Name _____

Date _____

Lesson 4—What Is the Scientific Method?

Use an Anticipation Guide

An anticipation guide can help you anticipate, or predict, what you will learn as you read.

- Look at the lesson titles and section titles for clues.
- Preview the Reading Focus Skill questions at the end of each section. Use what you know about the subject of each section to predict the answers.
- Read to find out whether your predictions were correct.

As you read each section, complete the anticipation guide below. Predict answers to each question and check to see if your predictions were correct.

Planning an Investigation		
Reading Focus Skill Question	**Prediction**	**Correct?**
What plan do scientists use to help them answer questions?	_____ _____ _____	
Conducting an Investigation		
Reading Focus Skill Question	**Prediction**	**Correct?**
_____ _____ _____	_____ _____ _____	

© Harcourt

Name _____

Date _____

Lesson 4—What Is the Scientific Method?

1. **Investigation Skill Practice–Compare**

Jefferson hypothesized that rabbits prefer carrots to celery. He experimented with 10 rabbits. His results supported his hypothesis. Diana hypothesized that parrots prefer peanuts to apples. She experimented with 2 parrots. The results did not support her hypothesis. Compare the two experiments, and write about what is different and what is alike.

_____

_____

_____

_____

2. **Focus Skill** **Reading Skill Practice–Main Idea and Details**

Read the selection. Underline the main idea. List at least three details about the main idea.

    All experiments end with the same step. The final step of an experiment is to draw a conclusion and communicate results. You can analyze your data to help you draw a conclusion. Next, you can decide whether the evidence supported your hypothesis. Finally, communicate your results in a written or oral report.

_____

_____

_____

© Harcourt

Use with *Getting Ready for Science.* (page 1 of 2) **Science Content Support** **CS 15**

Name _____

3. **Each statement below has a mistake. Rewrite the sentence correctly.**

   When you experiment, you use your senses to gather information.

   _____

   Conclusions can be used to conduct an investigation.

   _____

   A possible question is a hypothesis.

   _____

   You do not need to describe the steps when you plan an experiment.

   _____

   Your observations will be more accurate the fewer times you perform an experiment.

   _____

   _____

   Charts, tables, and rulers can help you communicate your results.

   _____

4. **Imagine you conducted an experiment and the results did not support your hypothesis. What could you do?**

   _____

   _____

   _____

© Harcourt

Name _____

Date _____

Lesson 1—Where Does Energy on Earth Come From?

Use Context Clues

Study the words and their definitions below. Then use context clues in each sentence to choose the word that belongs in the blank. Use all of the words once.

> **energy:** the ability to cause change
>
> **light:** a form of the sun's energy that you can see
>
> **heat:** the form of energy that warms objects
>
> **turbine:** a windmill that uses wind energy to produce electricity
>
> **oil:** a fuel that can be burned to create light and heat
>
> **lightning:** a natural type of electrical energy

1. Without _____ from the sun, none of Earth's plants would grow.

2. A _____ uses wind energy to produce electricity.

3. During the storm, a flash of _____ lit up the sky.

4. The sun's _____ prevents rivers and oceans from turning into ice.

5. Food supplies the _____ for runners to run in a marathon.

6. On cold days, a furnace burns _____ to heat your school or home.

© Harcourt

Use with Unit 1. **Science Content Support** **CS 17**

Name _____

Date _____

Lesson 1—Where Does Energy on Earth Come From?

Understand Vocabulary

Using a dictionary can help you learn new words that you find as you read.

- A dictionary shows all the meanings of a word and tells where the word came from.
- You can use a chart to list and organize unfamiliar words that you look up in a dictionary.

As you read this lesson, look up unfamiliar words in the dictionary. Add them to the chart below. Fill in each column to help you remember the word's meaning. Choose the meaning that matches the way the word is used in the lesson.

> **light** ('lit) *n* **1** something that makes vision possible
> **2** a source of light, such as a lamp   **3** daylight
> **4** general knowledge   **5** a source of fire, such as a match
> [from Old English *lēoht*]

Word	Syllables	Origin	Definition
light	LIGHT	Old English	Something that makes vision possible

© Harcourt

Lesson 1—Where Does Energy on Earth Come From?

1. **Investigation Skill Practice–Predict**

Suppose you placed two thermometers outside in a sunny spot. You placed the first thermometer face up on top of a sheet of white paper. The second thermometer was placed face up on top of a sheet of black paper. Predict what will happen when you record the temperatures of the two thermometers. Explain your prediction.

_____

_____

_____

2. **Focus Skill** **Reading Skill Practice–Main Idea and Details**

Read the selection. Underline the main idea. List three details about the main idea.

   Hazel is making breakfast for her family. Hazel's family will get energy from the things they eat and drink. Hazel makes pancakes from buckwheat, a plant. She slices fruit on top of the pancakes. The bananas, peaches, and apples grow on trees. She pours milk into glasses. The milk comes from cows, which eat plants. Plants and animals provide food for Hazel's family.

_____

_____

_____

© Harcourt

Name _____

3. In the chart below, write *plant* or *animal* in the second column to identify the source of each food. Then write three foods of your choice in the first column and complete the chart.

Food	Plant or Animal
peanut butter	
yogurt	
sausage	
broccoli	
orange juice	
tuna fish	

4. You already know that energy causes change. All of your actions use energy. Without energy, you would not be able to move, stretch, or grow. What are some ways you use energy after school?

_____

_____

_____

_____

Name _____

Date _____

Lesson 2—How Is Energy Stored and Used?

Explore Word Meanings

Write a word from the box to complete each sentence. Use glossary definitions and context clues in the sentences to help you choose the correct words.

fuel	battery	electricity	chemical	circuit

1. When you turn on your portable CD player, stored _____ energy changes to electricity.

2. A wire attached to the top and bottom of a battery forms a _____ .

3. Natural gas is a _____ formed from tiny living things in ancient oceans.

4. In a lamp bulb, _____ changes to light and heat.

5. When you put a new _____ in your toy car and turn it on, the toy car zooms across the floor.

© Harcourt

Name _____

Date _____

Lesson 2—How Is Energy Stored and Used?

Use a Graphic Organizer

A graphic organizer can help you make sense of the facts you read.

- Tables, charts, and webs are graphic organizers that can show main ideas and important details.
- A graphic organizer can help you classify and categorize information.

As you read this lesson, fill in each circle of the web below with facts from the lesson.

Stored Energy

Food

Fuel

Batteries

Gives people and animals energy to move and grow. Some of the energy from food helps keep our bodies warm.

© Harcourt

Name _____

Date _____

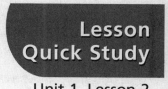

Lesson 2—How Is Energy Stored and Used?

1. | **Investigation Skill Practice–Draw Conclusions** |

Suppose that two identical balls are thrown against a wall. One ball simply bounces off the wall. The other ball causes the wall to crack. What can you conclude about the energy with which the balls were thrown?

_____

_____

_____

2. (Focus Skill) | **Reading Skill Practice–Sequence** |

Write the numbers 1–4 in the blanks to tell how energy from the sun changes to heat energy.

_____ The energy released from the logs warms the house.

_____ The logs are burned in a fireplace.

_____ The sun's energy is stored in a tree as it grows.

_____ The tree is cut down and chopped into logs.

© Harcourt

Name _____

3. Check (✓) the statements below that agree with the information found in the lesson.

_____ Ancient ocean organisms were pressed into oil when they died.

_____ Most plants store energy from insects in their leaves.

_____ Apple trees store the moon's energy in their fruit.

_____ Chemical energy is changed to electricity when a battery is used.

_____ Animals get energy by eating plants or other animals.

_____ Heat energy from the sun makes ice cubes melt.

4. **What things do you use that rely on batteries to supply energy? Explain how the energy helps you.**

_____

_____

_____

_____

© Harcourt

Name _____

Date _____

Lesson 3—How Does Energy Move?

Explore Word Meanings

Many words have several different meanings. Look at the meanings of the words below.

wave

    a. a disturbance that moves energy to other places

    b. a signal with an up-and-down or back-and-forth hand movement

vibrations

    c. a series of back-and-forth movements

    d. emotions that are sensed or felt

friction

    e. a force that slows the motion of objects that are touching

    f. disagreement or conflict between persons

Match the definitions of the three words to the underlined words in the sentences below.

_____ **1.** There was <u>friction</u> between the two students running for class president.

_____ **2.** Josh will <u>wave</u> goodbye when he leaves for San Diego.

_____ **3.** The singer's beautiful voice sent good <u>vibrations</u> through the audience.

_____ **4.** The huge <u>wave</u> caused the surfer to topple off his board.

_____ **5.** <u>Friction</u> slowed the ball down as it rolled across the carpet.

_____ **6.** The guitar strings <u>vibrated</u> as the girl plucked them.

Name _____

Date _____

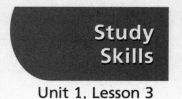

Lesson 3—How Does Energy Move?

Skim and Scan

Skimming and scanning are two ways to learn from what you read.

- To skim, quickly read the lesson title and the section titles. Look at the pictures and read the captions. Use this information to identify the main topics.
- To scan, look quickly through the text for specific details, such as key words or facts.

Before you read, skim the text to find the main idea of each section. Then look for key words. If you have questions, scan the text to find the answers.

SKIM	SCAN
Lesson: How Does Energy Move? **Main Idea:** Energy can be carried from place to place by waves, electricity, and moving objects. **Titles/Headings:** Waves Carry Energy, Sound Waves, Other Ways Energy Moves **Visuals:** _____ _____ _____ _____ _____	**Key Words and Facts** • wave • disturbance • energy • _____ _____ _____ _____ _____ _____ _____

© Harcourt

Name _____

Date _____

Lesson 3—How Does Energy Move?

1. ▆ Investigation Skill Practice–Use Models

Marc is using a model to study waves. His model is a rope that is tied to a doorknob. When he moves the free end of the rope up and down, waves move along the rope toward the doorknob. What kind of waves is Marc modeling? Explain your answer.

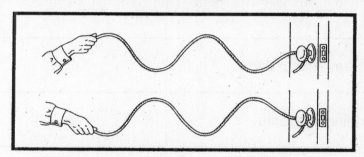

_____

_____

_____

2. ⭐(Focus Skill) ▆ Reading Skill Practice–Compare and Contrast

Read the selection. Tell how the waves are alike and different.

Maria stands at the end of a dock and makes waves on a calm lake. First, she tosses small rocks into the lake. The small rocks make small waves that have little energy. Then she tosses large rocks into the lake. The large rocks make big waves that have a lot of energy.

_____

_____

_____

© Harcourt

Name _____

Science Concepts

3. In the chart below, write *yes* or *no* in the second column to tell whether each form of energy travels in waves. Then write two more examples in the first column and complete the chart.

Movement of Energy	Yes or No
wind blowing on water	
the sun's light traveling to Earth	
electricity traveling to a lamp	
an earthquake traveling through the ground	
a bowling ball traveling to pins	
a skateboarder moving down a hill	

4. Energy can be transferred to a moving object when someone throws or kicks it. Which has more energy: the baseball that Lauren hit 12 feet or the soccer ball that Calvin kicked 4 feet? Explain why.

_____

_____

Name _____

Date _____

Lesson 4—What Is Matter?

Use a Graphic Organizer

Fill in the blank sections of the graphic organizer below.

Matter		
Three Forms of Matter	**Shape and Volume**	**One Example with Its Property**
solid	a kind of matter that has a definite shape and volume	
liquid		lemon juice: sour
	a kind of matter that has no definite shape or volume	Air: invisible

Explore Word Meanings

Write one or more sentences that use the words *matter, solid, liquid,* and *gas.*

_____

_____

_____

© Harcourt

Name _____

Date _____

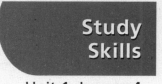

Lesson 4—What Is Matter?

Pose Questions

Asking questions as you read can help you understand what you are learning.

- Form questions as you read. Think about *why* and *how* events happened and *how* events and ideas are related.
- Use the questions to guide your reading. Look for the answers as you read.

As you read the lesson, fill in the chart below. Write down any questions you have about ideas, details, vocabulary words, or other information. Then read on to look for the answers.

What Is Matter?	
Questions	Answers
What is volume?	Volume is the amount of space something takes up.
What are three forms of matter?	_____ _____
_____ _____	_____ _____
_____ _____	_____ _____

Name _____

Date _____

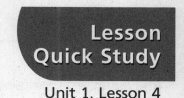

Lesson 4—What Is Matter?

1. **Investigation Skill Practice—Predict**

Suppose you pour 300 mL of water into a narrow container. You mark the water level with a marker. Next, you will pour the water into a wide container. Predict whether the water level in the wide container will be higher or lower than the level in the narrow container. Explain your prediction.

_____

_____

_____

2. **(Focus Skill) Reading Skill Practice—Main Idea and Details**

Read the selection. Underline the main idea. List at least three details about the main idea.

    You can use your senses to observe different properties of matter. You can use your sense of touch to feel the rough texture of burlap. You can use your sense of taste to sample salty peanuts. You can use your sense of smell to sniff sweet roses. You can use your sense of sight to see orange oak leaves.

_____

_____

_____

© Harcourt

Name _____

3. Draw a line from the type of matter to the tool that can be used to measure one of its properties. Use each tool twice.

Matter	Tool
length of a pencil	measuring cup
milk	
5 pennies	balance
10 marbles	
olive oil	ruler
width of wooden shelf	

4. Matter has different characteristics you can observe by using your senses. Imagine that you are watching a movie in a theater. What are some characteristics of the matter that surrounds you?

_____

_____

_____

Name _____

Date _____

Lesson 5—What Causes Matter to Change State?

Use Context Clues

Study the words and their definitions below. Then use context clues in each sentence to choose the word that belongs in the blank. Use all of the words once.

> **melt:** to change from a solid to a liquid
> **boil:** to change from a liquid to a gas
> **condense:** to change from a gas to a liquid
> **freeze:** to change from a liquid to a solid
> **evaporate:** to change from a liquid to a gas

1. When Jenny stood in the sun after swimming, the water from her bathing suit began to _____ .

2. After being caught in traffic, Mom rushed home to _____ the melting ice cream.

3. I heated the butter in a pan on the stove so that it would _____.

4. In five minutes, the water in the teapot began to _____ and make a hissing sound as the steam escaped.

5. Water begins to _____ on a cold can of juice when you drink it outside on a hot summer day.

Name _____

Date _____

Lesson 5—What Causes Matter to Change State?

Write to Learn

Writing about what you read can help you understand and remember information.

- Many students write about their reading in learning logs.
- Writing about the text leads you to think about it.
- Writing your reactions to the text makes it more meaningful to you.

As you read the lesson, pay attention to new and important information. Keep track of the information by completing the learning log below.

Water Changes

What I Learned	My Thoughts
When ice is heated, it can melt. When water is heated, it evaporates.	_____ _____ _____ _____
_____ _____ _____	_____ _____ _____
_____ _____ _____	_____ _____ _____

Use with Unit 1.

Name _____

Date _____

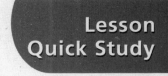

Lesson 5—What Causes Matter to Change State?

1. **Investigation Skill Practice–Communicate**

Observe the picture below. Draw a picture in the empty box to communicate how the ice will change after one hour. Write a few sentences to explain the process.

_____

_____

2. **Focus Skill** **Reading Skill Practice–Cause and Effect**

Read the selection. Then describe a cause and an effect of freezing.

    Jasmine wanted to make ice pops for her friends. She poured grape juice into plastic molds and snapped the holders into the molds. Then Jasmine put the molds with the grape juice into the freezer. In a few hours, the cold air caused the juice to freeze. Jasmine popped the frozen juice out of the molds and enjoyed the grape ice pops with her friends.

_____

_____

Name _____

3. In the chart below, write *melt, boil, condense, freeze,* or *evaporate* in the second column to identify the change of state. Use each word once.

Change of State

a chocolate bar turns to liquid in your jacket pocket	
raindrops disappear on an umbrella drying in the sun	
leftover soup in the freezer changes from liquid to solid	
a pot of water bubbles on the stove	
water vapor becomes droplets of liquid water on the outside of a glass of iced tea	

4. What states of water can you find in nature? Give an example of each state.

_____

_____

_____

© Harcourt

Name _____

Date _____

Lesson 6—What Are Physical and Chemical Changes?

Explore Word Meanings

Write a word from the box to complete each sentence. Use glossary definitions and context clues in the sentences to help you choose the correct words.

react	mixture	substance	dissolve	separate

1. When you stir a spoonful of salt into hot water, the salt will _____ and disappear.

2. Thomas made a _____ of oil and vinegar to pour over the salad.

3. Oxygen in the air will _____ with iron to form an orange substance known as rust.

4. You can use a magnet to _____ iron filings from sawdust.

5. In some chemical changes, when you mix one _____ with another, bubbles of gas are formed.

Name _____

Date _____

Lesson 6—What Are Physical and Chemical Changes?

Use a K-W-L Chart

A K-W-L chart can help you focus on what you already know about a topic and what you want to learn about it.

- Use the *K* column to list what you already know about physical and chemical changes.
- Use the *W* column to list what you want to learn about physical and chemical changes.
- Use the *L* column to list what you have learned about the topic from your reading.

Complete the K-W-L chart as you read this lesson.

Physical and Chemical Changes		
What I Know	**What I Want to Learn**	**What I Learned**
• When my mom bakes muffins, the mixture changes color and texture, and it rises.	• What causes the mixture to rise? • What happened to the original ingredients?	
• _____ _____ • _____ _____ • _____ _____	• _____ _____ • _____ _____ • _____ _____	• _____ _____ • _____ _____ • _____ _____

Name _____

Date _____

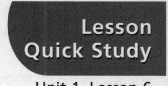

Lesson 6—What Are Physical and Chemical Changes?

1. **Investigation Skill Practice—Compare**

Observe the pictures below. Compare the way in which the clay and the wood change.

_____

_____

_____

2. **(Focus Skill)** **Reading Skill Practice—Main Idea and Details**

Read the selection. Underline the main idea. List at least two details about the main idea.

    Matter can go through both physical and chemical changes. Suppose you want to build a treehouse of wood. After several days of hard work, the pile of wood you started out with is now a treehouse! The change you observed—from individual boards to a treehouse—is a physical change. Then a terrible disaster happens. Lightning strikes your treehouse, and fire destroys it. Your treehouse has become smoke and ashes. This change in the wood is a chemical change.

_____

_____

_____

© Harcourt

Name _____

3. Check (✓) the statements below that agree with the information found in the lesson.

   _____ A slice of apple reacts with oxygen and turns brown.

   _____ After rust forms, it can be changed back into iron.

   _____ If you let a glass of salt water evaporate, the salt is left in the glass.

   _____ When you stir sugar into a pitcher of lemonade, the sugar dissolves.

   _____ You cannot separate a mixture of sand, marbles, and water.

   _____ A snowball melting in your mitten undergoes a chemical change.

   _____ During a chemical change, light and heat energy can be released.

4. Mrs. Garcia has lettuce, celery, tomatoes, oil, and salt. Describe a solution and a mixture she can make using all these ingredients.

   _____

   _____

   _____

   _____

© Harcourt

Name _____

Date _____

Unit 1, Lesson 7

Lesson 7—What Are Atoms and Elements?

Use a Graphic Organizer

Fill in the blank sections of the graphic organizer below.

Words to Describe Matter	Definition
	the smallest particle of matter that still has the properties of that matter
element	
	a chart showing the names and symbols of more than 100 elements
particle	
	one or more letters that stand for each element in the periodic table
pure substance	

© Harcourt

Use with Unit 1.

Science Content Support CS 41

Lesson 7—What Are Atoms and Elements?

Make an Outline

An outline is a good way to record main ideas and details.

- Topics in an outline are shown by Roman numerals.
- Main ideas about each topic are shown by capital letters.
- Details about each main idea are identified by numbers.

As you read this lesson, complete the outline below.

What Are Atoms and Elements?

I. Atoms

    A. Greek thinkers first proposed the idea

    B. Smallest particles of matter

        1. Millions of times smaller than sand grains

        2. Can barely be seen with strongest microscopes

II. Elements

    A. Building blocks of matter

    _____

    _____

III. Combining Elements

    A. Can be combined in different ways

    _____

    _____

    B. All living things are mostly combinations of a few elements

    _____

© Harcourt

Name _____

Date _____

Lesson 7—What Are Atoms and Elements?

1. Investigation Skill Practice–Use Models

Jenna made a model to show tiny particles of matter. First, she measured a tablespoon of coffee beans. Then, she used a coffee grinder to grind the beans into tiny specks. What do the tiny specks of coffee beans represent? Tell how you know.

_____

_____

_____

2. (Focus Skill) Reading Skill Practice–Main Idea and Details

Read the selection. Underline the main idea. List at least two details about the main idea.

Elements are named by the scientists who discover them. This tradition started in the 1600s when scientists began to separate elements from each other. Some elements are named after places, such as Berkeley, California (Berkelium, Bk). Other elements are named after famous scientists, such as Albert Einstein (Einsteinium, Es). Still other elements are named after characters in mythology, such as Promethius (Promethium, Pm).

_____

_____

_____

_____

© Harcourt

Name _____

3. Write *True* or *False* in the blanks. If the statement is
false, write a new statement on the line.

_____ If you could split a gold atom in half, it would still
be gold.

_____

_____ The smallest piece of aluminum you can have is an
aluminum atom.

_____

_____ Early Greeks thought that four elements—earth, air,
fire, and water—made up matter.

_____

_____ There are more than 1,000 elements in the periodic
table.

_____

_____ Elements have several kinds of atoms.

_____

_____ Neither water nor salt are found on the periodic table.

_____

_____ Air appears on the periodic table.

_____

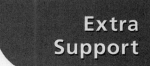
Periodic Table

Key

6	— symbol
C	
Carbon	— name

1																	18
1 **H** Hydrogen	2											13	14	15	16	17	2 **He** Helium
3 **Li** Lithium	4 **Be** Beryllium											5 **B** Boron	6 **C** Carbon	7 **N** Nitrogen	8 **O** Oxygen	9 **F** Fluorine	10 **Ne** Neon
11 **Na** Sodium	12 **Mg** Magnesium	3	4	5	6	7	8	9	10	11	12	13 **Al** Aluminum	14 **Si** Silicon	15 **P** Phosphorus	16 **S** Sulfur	17 **Cl** Chlorine	18 **Ar** Argon
19 **K** Potassium	20 **Ca** Calcium	21 **Sc** Scandium	22 **Ti** Titanium	23 **V** Vanadium	24 **Cr** Chromium	25 **Mn** Manganese	26 **Fe** Iron	27 **Co** Cobalt	28 **Ni** Nickel	29 **Cu** Copper	30 **Zn** Zinc	31 **Ga** Gallium	32 **Ge** Germanium	33 **As** Arsenic	34 **Se** Selenium	35 **Br** Bromine	36 **Kr** Krypton
37 **Rb** Rubidium	38 **Sr** Strontium	39 **Y** Yttrium	40 **Zr** Zirconium	41 **Nb** Niobium	42 **Mo** Molybdenum	43 **Tc** Technetium	44 **Ru** Ruthenium	45 **Rh** Rhodium	46 **Pd** Palladium	47 **Ag** Silver	48 **Cd** Cadmium	49 **In** Indium	50 **Sn** Tin	51 **Sb** Antimony	52 **Te** Tellurium	53 **I** Iodine	54 **Xe** Xenon
55 **Cs** Cesium	56 **Ba** Barium	57 **La** Lanthanum	72 **Hf** Hafnium	73 **Ta** Tantalum	74 **W** Tungsten	75 **Re** Rhenium	76 **Os** Osmium	77 **Ir** Iridium	78 **Pt** Platinum	79 **Au** Gold	80 **Hg** Mercury	81 **Tl** Thallium	82 **Pb** Lead	83 **Bi** Bismuth	84 **Po** Polonium	85 **At** Astatine	86 **Rn** Radon
87 **Fr** Francium	88 **Ra** Radium	89 **Ac** Actinium	104 **Rf** Rutherfordium	105 **Db** Dubnium	106 **Sg** Seaborgium	107 **Bh** Bohrium	108 **Hs** Hassium	109 **Mt** Meitnerium	110 **Ds** Darmstadtium	111 **Rg** Roentgenium	112 **Uub** Ununbium	113 **Uut** Ununtrium	114 **Uuq** Ununquadium	115 **Uup** Ununpentium	116 **Uuh** Ununhexium		

58 **Ce** Cerium	59 **Pr** Praseodymium	60 **Nd** Neodymium	61 **Pm** Promethium	62 **Sm** Samarium	63 **Eu** Europium	64 **Gd** Gadolinium	65 **Tb** Terbium	66 **Dy** Dysprosium	67 **Ho** Holmium	68 **Er** Erbium	69 **Tm** Thulium	70 **Yb** Ytterbium	71 **Lu** Lutetium
90 **Th** Thorium	91 **Pa** Protactinium	92 **U** Uranium	93 **Np** Neptunium	94 **Pu** Plutonium	95 **Am** Americium	96 **Cm** Curium	97 **Bk** Berkelium	98 **Cf** Californium	99 **Es** Einsteinium	100 **Fm** Fermium	101 **Md** Mendelevium	102 **No** Nobelium	103 **Lr** Lawrencium

Name _____

Use the periodic table on the previous page to answer the following questions.

1. What is the first element at the top left of the table?

_____

2. What element is to the right of carbon (C)?

_____

3. Elements can combine to make thousands of substances. Fill in the chart below with the names and symbols of the elements that make up each substance named in the left column.

Substance	Elements
salt (NaCl)	sodium (Na) + chlorine (Cl)
ammonia (NH_3)	
rust (Fe_2O_3)	
table sugar ($C_{12}H_{22}O_{11}$)	

4. List four elements you are familiar with. Next to each element, list an example of something that contains that element.

Element	Example
gold	earrings

© Harcourt

Name _____

Date _____

Lesson 1—How Does Light Travel?

Use Context Clues

**Read the sentences. Think about the meaning of each
underlined word. Write the meaning on the line.**

1. When light bounces off your body, travels to a mirror, and bounces
back to your eyes, you can see your <u>reflection</u>.

_____

2. On a sunny day at noon, your <u>shadow</u> is a small, dark area that is
close to your feet.

_____

_____

3. <u>Light</u> energy allows you to see colors and objects.

_____

4. As burning wood changes to ashes and hot gases, heat and light
<u>energy</u> are released. Also, the <u>energy</u> in food enables your body to
move and grow.

_____

5. The <u>source</u> of light in a reading lamp is a light bulb.

_____

© Harcourt

Name _____

Date _____

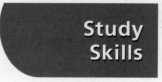

Lesson 1—How Does Light Travel?

Make an Outline

An outline is a good way to record main ideas and details.

- Topics in an outline are shown by Roman numerals.
- Main ideas about each topic are shown by capital letters.
- Details about each main idea are identified by numbers.

As you read this lesson, remember to pay attention to the topics, main ideas, and details. Complete the outline below.

How Does Light Travel?
- I. Light Energy
  - A. What we can see
    - 1. Objects that reflect light
    - 2. _____
    - 3. Things that produce light
  - B. Nature of light
    - 1. _____
    - 2. Can be seen
    - 3. _____
    - 4. _____
- II. Shadows
  - A. Dark areas formed because path of light is blocked
    - 1. _____
    - 2. _____
- III. _____
  - A. How Light Reflects
    - 1. Bounces off objects and surfaces
    - 2. _____
    - 3. _____

Name _____

Date _____

Lesson 1—How Does Light Travel?

1. **Investigation Skill Practice–Compare**

Observe the shadows in the diagrams below. Compare the two shadows, and explain how they differ.

2 P.M.   **6 P.M.**

_____

_____

2. (Focus Skill) **Reading Skill Practice–Main Idea and Details**

Read the selection. Underline the main idea. List at least three details about the main idea.

    Douglas fills a flowerpot with dirt and pokes a hole in the dirt. Then he gently places a sunflower seedling in the hole, pats dirt around the roots, and sprinkles it with water. He puts the flowerpot on a sunny windowsill. Douglas uses a ruler to measure how much the seedling grows each day. The seedling grows more on sunny days. Soon Douglas needs to transplant the seedling to a bigger flowerpot. Light energy from the sun has caused the sunflower seedling to grow.

© Harcourt

_____

_____

_____

Name _____

3. In the chart below, write *yes* or *no* in the second column
   to tell whether you can or cannot see your reflection in
   the materials or things listed. Then write the names of two
   materials or things in the first column and complete the
   chart.

Material	Yes or No
pond	
chalkboard	
wood door	
aluminum foil	
metal spoon	
carpet	

4. Charlotte stands between a wall and a light bulb. As she
   moves her arms, her shadow on the wall moves in the
   same way. Based on what you know about light, what
   produces Charlotte's shadow?

_____

_____

_____

_____

© Harcourt

Name _____

Date _____

Lesson 2—What Causes Color?

Explore Word Meanings

Many words have several different meanings. Look at the meanings of the words below.

absorb
- **a.** to take something in
- **b.** to occupy someone's full attention

reflect
- **c.** to bounce off
- **d.** to think seriously

rainbow
- **e.** an arc of colors formed when a prism bends sunlight
- **f.** a mixed collection

Match the above definitions of the three words to the underlined words in the sentences below.

_____ **1.** Josh has a <u>rainbow</u> of tropical fish in his aquarium.

_____ **2.** A banana <u>absorbs</u> every color of light except yellow.

_____ **3.** In white light, a tangerine <u>reflects</u> orange light.

_____ **4.** The school play <u>absorbed</u> all of Mandy's spare time.

_____ **5.** When the sun came out after the rain storm, Tim saw a <u>rainbow</u>.

_____ **6.** Juanita <u>reflected</u> on the math problem before writing her answer.

© Harcourt

Lesson 2—What Causes Color?

Use Visuals

Visuals can help you better understand and remember what you read.

- Photographs, illustrations, diagrams, charts, and maps are different kinds of visuals. Many visuals have titles, captions, or labels that help readers understand what is shown.
- Visuals often present information from the reading, but in a different way. They may also add information.

As you read this lesson, look closely at the visuals and the words that go with them. Answer the questions in the checklist.

	Checklist for Visuals
✔	What kind of visual is shown? a diagram
✔	What does the visual show? the order of colors in a rainbow
	What does the visual tell you about the topic? _____ _____ _____
	How does the visual help you understand what you are reading? _____ _____ _____

© Harcourt

Lesson 2—What Causes Color?

1. | Investigation Skill Practice—Predict |

In the Investigate, you used what you already knew about light and color to predict the outcome of the activity. Suppose you use a sheet of green paper for the activity. What do you predict the light will look like on the green paper? Explain.

_____

_____

_____

2. (Focus Skill) | Reading Skill Practice—Cause and Effect |

Draw and complete this graphic organizer.

Cause	Effect

| White light enters a prism. | → | White light is _____ into its component colors. |

| White light shines on an apple. | → | Every color of light except red is _____. |

| Light strikes a mirror. | → | The mirror _____ most of the light that strikes it. |

Name _____

3. Check (✓) the statements below that agree with the information found in the lesson.

_____ The order of colors in a rainbow is red, orange, yellow, green, blue, and violet.

_____ A lemon reflects every color of light except yellow.

_____ Black objects absorb more light than white objects.

_____ Sunshine and water droplets are needed to form a rainbow.

_____ Shiny objects absorb more light than dull objects.

4. **You already know that light has many colors. When white light hits an object, we do not see many colors. We see the color that the object reflects. The other colors are absorbed. Explain what happens when white light shines on a green apple.**

_____

_____

_____

© Harcourt

Name _____

Date _____

Lesson 3—How Do You See Objects?

Use Context Clues

Study the words and their definitions below. Then use context clues in each sentence to choose the word that belongs in the blank. Use all of the words once.

<u>transparent</u>:   an object that lets most light pass through it

<u>retina</u>:   the back inside surface of the eye, the part of the eye that is connected by a nerve to the brain

<u>opaque</u>:   an object that does not let any light pass through it

<u>iris</u>:   the colored part of the eye that controls how much light enters

<u>translucent</u>:   an object that lets some light pass through it

<u>pupil</u>:   a black spot in the center of the eye through which light passes

1. The frosted glass of winter holiday lights is _____.

2. More light passes into the eye when the _____ is larger.

3. Light cannot travel through _____ objects, such as desks and wooden doors.

4. In bright sunshine, the _____ makes the pupil smaller.

5. The clear glass in your bedroom window is _____.

6. A nerve carries information from the _____ to your brain.

© Harcourt

Name _____

Date _____

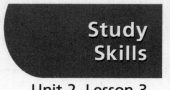

Lesson 3—How Do You See Objects?

Use A K-W-L Chart

A K-W-L chart can help you focus on what you already know about a topic and what you want to learn about it.

- Use the K column to list what you know about a topic.
- Use the W column to list what you want to know about the topic.
- Use the L column to list what you have learned about the topic after reading.

As you read this lesson, complete the K-W-L chart below.

Light Interactions		
What I <u>K</u>now	**What I <u>W</u>ant to Know**	**What I <u>L</u>earned**
• I can see through the clear glass of a window.	• How much light passes through transparent objects?	• _____ _____ _____ _____
• _____ _____ _____ _____	• _____ _____ _____ _____	• _____ _____ _____ _____
• _____ _____ _____ _____	• _____ _____ _____ _____	• _____ _____ _____ _____
• _____ _____ _____	• _____ _____ _____	• _____ _____ _____

© Harcourt

Lesson 3—How Do You See Objects?

1. **Investigation Skill Practice–Communicate**

Imagine you and a friend built a fort using the three windows shown below. Using what you know about light, explain what you would see inside the fort as your friend stood outside and shined a flashlight on each window.

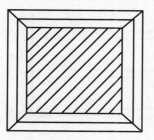
translucent

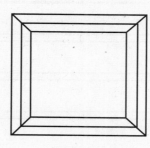
transparent

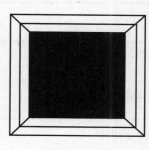
opaque

_____

_____

_____

_____

2. **Focus Skill** **Reading Skill Practice–Sequence**

Number the following steps from 1 to 5 to tell how you see a book.

_____ Light shines on the retina.

_____ The brain tells you that you see a book.

_____ A nerve carries information to the brain.

_____ Light passes through the lens, which bends the light.

_____ Light bounces off the book and passes through the pupil.

© Harcourt

Name _____

3. In the chart below, write *transparent*, *translucent*, or
   opaque in the second column to tell if the object lets all
   light through, some light through, or no light through.
   Then write the names of three other objects or materials
   in the first column and complete the chart.

Object	Transparent, Translucent, or Opaque
eyeglasses	
comic book	
frosted mug	
wax paper	
mirror	
camera lens	

4. You already know that the iris, or colored part of the eye,
   controls how much light enters the eye. For example, on
   a bright, sunny day, the iris makes the pupil small so that
   only a little light can get through. Describe the size of
   the iris at dusk, when light is dim.

_____

_____

© Harcourt

Name _____

Date _____

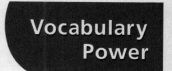
Lesson 1—How Do Plant Parts Help Plants?

Word Families

Fill in the blanks with the correct form of the word.

1. survival/survive

<u>Noun</u>: The _____ of desert plants depends on their ability to save water.

<u>Verb</u>: Insects help a Venus' flytrap _____.

2. adapt/adaptation

<u>Noun</u>: Thorns and spines are examples of an _____.

<u>Verb</u>: Some vines _____ by wrapping around other plants and climbing upward.

3. reproduction/reproduce

<u>Noun</u>: During _____, dogs have puppies.

<u>Verb</u>: A sunflower uses seeds to _____.

Explore Word Meanings

Think about the meaning of the underlined words. Answer each question.

4. Your <u>traits</u> are what you look like. What are some of your *traits*?

_____

5. The deer was able to <u>survive</u> the fire. What do you think the deer did to *survive* the fire?

_____

© Harcourt

Name _____

Date _____

Lesson 1—How Do Plant Parts Help Plants?

Understand Vocabulary

Using a dictionary can help you learn new words that you find as you read.

- A dictionary shows all the meanings of a word and tells where the word came from.
- You can use a chart to list and organize unfamiliar words that you look up in a dictionary.

As you read the lesson, look up unfamiliar words in the dictionary. Add them to the chart below. Fill in each column to help you remember the word's meaning.

trait (trát) *n.* **1.** A feature that sets one person or thing off from another. **2.** An inherited characteristic. (from early French *trait,* "the act of pulling")

Word	Syllables	Origin	Definition
Trait	TRAIT	early French	A feature that sets one person or thing off from another

© Harcourt

Lesson 1—How Do Plant Parts Help Plants?

1. Investigation Skill Practice—Compare

Jackson went for a walk in the woods. He drew pictures of plants that have adaptations that help them reproduce far away from the parent plant. Describe the adaptation of each seed.

_____

_____

_____

_____

_____

_____

2. (Focus Skill) Reading Skill Practice—Main Idea and Details

Read the selection. Underline the main idea. List at least three details about the main idea.

Plants make seeds using pollen from other flowers. Some plants use smell, color, and shape to attract animals to their pollen. For example, hummingbirds are attracted to nectar, a sweet-smelling liquid produced by flowers. Red flowering plants also lure hummingbirds. Long-necked flowers attract hummingbirds, which have long, slender beaks.

_____

_____

_____

_____

Name _____

3. In the chart below, write *yes* on the right if the
adaptation helps plants reproduce. Write *no* on the right
if the adaptation does not help plants reproduce. Write
two adaptations of your choice on the left and complete
the chart.

Adaptation	Helps Plants Reproduce?
sticky seeds	
large leaves	
long roots	
seeds float in water	
thick stems	
seeds carried by wind	
color	
sweet scent	

4. Plants have adaptations that help them grow, survive,
and reproduce. Explain why a desert plant, such as the
cactus, cannot survive in a pond.

_____

_____

_____

_____

© Harcourt

Name _____

Date _____

Lesson 2—How Do Animal Adaptations Help Animals?

Graphic Organizer

Fill in the blank sections of the graphic organizer below.

Animal Adaptations		
Ways that Animals Survive	**Definition**	**Animal Examples**
migrate		whale, monarch butterfly
change colors	to blend in by switching colors to match the surroundings	
	to spend the winter in deep sleep	chipmunk, bat
quills		porcupine
claw	a sharp nail or a pincher	
	a pocket of skin in which a baby lives	kangaroo

Lesson 2—How Do Animal Adaptations Help Animals?

Skim and Scan

Skimming and scanning are two ways to learn from what you read.

- To skim, quickly read the lesson title and the section titles. Look at the pictures and read the captions. Use this information to identify the main topics.
- To scan, look quickly through the text for specific details, such as key words or facts.

Before you read, skim the text to find the main idea of each section. Then look for key words. If you have questions, scan the text to find the answers. Fill in the chart below.

SKIM	SCAN
Lesson: How Do Animal Adaptations Help Animals? **Main Idea:** Animals have adaptations that help them grow, survive, and reproduce. **Titles/Headings:** Adaptations for Survival, _____ _____ _____ **Visuals:** _____ _____ _____ _____	**Key Words and Facts** • adaptation • migrate • hibernate • _____ • _____ • _____ • _____ • _____ • _____ • _____

Name _____

Date _____

Lesson 2—How Do Animal Adaptations Help Animals?

1. **Investigation Skill Practice–Experiment**

Imagine that you want to study your pet birds as part of your science fair project. You think that birds with larger beaks can crack open seeds faster than birds with smaller beaks. Explain how you would design an experiment to test your hypothesis.

_____

_____

_____

_____

2. **Focus Skill** **Reading Skill Practice–Cause and Effect**

Read the selection. Underline what causes the Arctic fox to change its color. List two effects of this color change.

The Arctic fox has a coat of gray-brown fur in warm weather and a thick coat of white fur in cold weather. Since the Arctic fox does not hibernate, its white fur camouflages it against the winter landscape of snow and ice. The white fur prevents predators from easily finding the fox and turning it into a meal. The thick white fur also conserves heat in the cold Arctic winter.

_____

_____

© Harcourt

Name _____

Science Concepts

3. **Check (✓) the statements below that agree with the information found in the lesson.**

   _____ Sharp claws and beaks help eagles grab and eat food.

   _____ Blubber keeps animals cold in warm weather.

   _____ When animals hibernate, they live off their body fat.

   _____ Hearts of hibernating animals beat faster than usual.

   _____ Gray whales swim south to warmer waters in the fall.

   _____ Some animals keep safe by looking like other animals.

   _____ Animals need to be taught how to hibernate and migrate.

   _____ A chameleon is able to change colors.

4. **Why do you think some animals that live in the north hibernate, while the animals that live in the south do not?**

   _____

   _____

   _____

   _____

   _____

   _____

© Harcourt

Name _____

Date _____

Lesson 2—How Do Animal Adaptations Help Animals?

Animal Adaptations

A group of scientists had a question about animal adaptations. The scientists wanted to design an experiment to test an adaptation that would help an animal to survive. They headed to Africa to collect data.

The scientists planned an experiment that focused on the tallest land animals: giraffes. They counted how many pounds of acacia leaves that 10 giraffes ate in a week. They also measured how tall each giraffe was. They recorded their data in the table below.

Giraffe Data		
Giraffe Name	**Height (in feet)**	**Pounds of Leaves Consumed in 7 Days**
Alpha	14	450
Beta	10	350
Gamma	17	500
Delta	11	375
Epsilon	15	460
Zeta	16	475
Eta	12	400
Theta	14	430
Iota	18	525
Kappa	13	420

© Harcourt

Name _____

Date _____

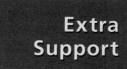

A. How would you interpret the data that the scientists gathered? Compare measurements to draw a logical conclusion.

_____

_____

_____

_____

_____

_____

B. Scientists use facts to back up their observations. They do not rely on claims or opinions. Read the statements below. Put a check next to the statements that are supported by evidence from the experiment.

_____ Giraffes of different subspecies have different spot patterns on their coats.

_____ Giraffes are the most exciting animals in Africa.

_____ Gamma was one of the tallest giraffes studied.

_____ There was a 175 pound difference in pounds of leaves consumed between the tallest and the shortest giraffes.

_____ Giraffes are not currently endangered animals.

_____ Beta, the shortest giraffe, ate 350 pounds of leaves.

_____ Iota, the tallest giraffe, was 18 feet in height.

_____ A giraffe's neck weighs about 60 pounds.

_____ When giraffes aren't eating leaves, they are sleeping.

_____ Shorter giraffes usually eat fewer leaves.

© Harcourt

Name _____

Date _____

Lesson 3—What Lives in Different Environments?

Explore Word Meanings

Think about the meanings of the underlined words. Then write your answer to each question.

1. A grizzly bear's <u>environment</u> includes trees, streams, other animals, and the food the bear eats. What is an *environment*?

_____

2. A <u>habitat</u> is the place in an environment where a plant or animal lives. What is an ant's *habitat*?

_____

3. <u>Climate</u> is what the weather is like in a place. What is the *climate* like where you live?

_____

Analogies

An analogy is made of two pairs of words. The words in each pair are related to each other in the same way. Choose a word from the box that completes each analogy.

wetland	forest	grassland

4. Fish is to river as tree is to _____.

5. Cactus is to desert as grass is to _____.

6. Whale is to ocean as frog is to _____.

© Harcourt

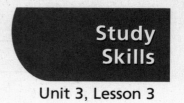

Lesson 3—What Lives in Different Environments?

Pose Questions

Asking questions as you read can help you understand what you are learning.

- Form questions as you read. Think about why and how events happened and how events and ideas are related.
- Use the questions to guide your reading. Look for the answers as you read.

As you read, fill in the chart below. Write down any questions you have about ideas, details, vocabulary, experiments, and other information discussed in the lesson. Then read on to look for the answers.

Oceans	
Questions	**Answers**
What is found in an ocean environment?	Water, sunlight, seaweed, and fish are found in an ocean environment.
_____ _____	_____ _____
_____ _____	_____ _____
_____ _____	_____ _____

© Harcourt

Name _____

Date _____

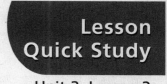

Lesson 3—What Lives in Different Environments?

1. **Investigation Skill Practice–Draw Conclusions**

Use what you have learned to draw conclusions about the plants and animals listed below. What habitat does each of the plants and animals live in? How do you know?

     Arctic hare   cactus   squirrel

_____

_____

_____

_____

2. **Focus Skill** **Reading Skill Practice–Compare and Contrast**

Read the selection. Compare and contrast the dolphin and the squid.

    Dolphins and squids live in the ocean. Both are fast swimmers and both eat fish. Dolphins come to the surface to breathe oxygen. Squids take in oxygen from the water. Dolphins give birth to live young and produce their own milk to feed their young. Squids lay eggs.

_____

_____

_____

_____

© Harcourt

Name _____

3. **Match the animal to its habitat. Draw a line from the animal to the habitat where it lives. Use each habitat twice.**

Animal	Habitat
reindeer	forest
rabbit	
zebra	tundra
duck	
deer	grassland
camel	
bison	wetland
turtle	
polar bear	desert
ringtail	

4. **Read the paragraph below and answer these questions. What do you think happened to the frogs? Why do you think it happened?**

   A large number of frogs lived in the wetlands at the edge of town. Then the wetlands were drained to build a shopping mall. The trees were cut for shops and the streams were built over. Now there are no frogs in this area.

_____

_____

_____

_____

_____

© Harcourt

Name _____

Date _____

Lesson 3—What Lives in Different Environments?

Six Different Environments

In the boxes below, describe each environment. Then draw a plant or animal that lives in that environment. Explain one adaptation that the plant or animal has that helps it survive in its environment.

Desert	
_____ _____ _____ _____	

Ocean	
_____ _____ _____ _____	

© Harcourt

Name _____

Wetland	
_____ _____ _____	

Forest	
_____ _____ _____	

Tundra	
_____ _____ _____	

Grassland	
_____ _____ _____	

Name _____

Date _____

Lesson 4—How Do Living Things Change Environments?

Graphic Organizer

Fill in the blank sections of the graphic organizer below.

Pollution		
Types of Pollution	**An Example of Each**	**What It Harms**
air pollution		damages peoples' lungs
	oil spilled from a tanker	kills fish and birds
land pollution	trash and chemicals thrown on the ground	

Name _____

Date _____

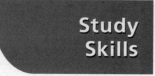

Lesson 4—How Do Living Things Change Environments?

Organize Information

A graphic organizer can help you make sense of the facts you read.

- Tables, charts, and webs are graphic organizers that can show main ideas and important details.
- A graphic organizer can help you classify and categorize information. It can also help you understand the relationship between the subject of the chapter and each lesson.

As you read this lesson, fill in each circle of the web below with facts from the lesson.

How Living Things Change Environments

People Harm the Environment	People Help the Environment	Living Things Change Their Environment
_____	_____	_____
_____	_____	_____
_____	_____	_____
_____	_____	_____
_____	_____	_____
_____	_____	_____

© Harcourt

Name _____

Date _____

Lesson 4—How Do Living Things Change Environments?

1. **Investigation Skill Practice—Compare**

Suppose a lake has been polluted with trash. Another lake has been polluted with oil from a factory. Compare how you think the animals in the polluted lakes are affected.

_____

_____

_____

_____

_____

2. **(Focus Skill)** **Reading Skill Practice—Sequence**

Put the following events about beaver dams in the correct sequence. Number the steps 1 to 5.

_____ Beavers add logs and branches to the dam.

_____ The pond slowly fills with dirt, and the beavers move to a new spot.

_____ Beavers start building a small dam on a stream.

_____ Beavers hide in the deep water, away from predators.

_____ The water slows and becomes deeper.

© Harcourt

Name _____

3. In the chart below, write how people can help the environment when something harmful happens.

Harmful Action	Helpful Action
trees are cut down to build houses	
cars and trucks cause air pollution	
factories dump chemicals into water	
farmers spray crops with chemicals to kill insect pests	
people throw out 4 pounds of trash every day	

4. Suppose a tree falls in a forest. How will this change affect living things in the forest? Is the change helpful, harmful, or both helpful and harmful? Explain.

_____

_____

_____

_____

Name _____

Date _____

Lesson 5—How Do Changes to Environments Affect Living Things?

Explore Word Meanings

Many words have more than one meaning. Look at the meanings of the words below.

drought

    a. a long period of time with very little rain

    b. a long lack of something important

balance

    c. not too many or not too few of any living thing

    d. an instrument used for measuring mass

flood

    e. a period of time when water covers land that is usually dry

    f. to send a lot of calls or letters to one place

Match the above definitions to the underlined words in the sentences below.

_____ **1.** Three straight days of rain caused the <u>flood</u>.

_____ **2.** The new plants upset the <u>balance</u> of the pond.

_____ **3.** All of the flowers I planted died during the <u>drought</u>.

_____ **4.** Use the <u>balance</u> to find the baby rabbit's mass.

_____ **5.** There was a <u>drought</u> of good ideas in the classroom.

_____ **6.** They received a <u>flood</u> of mail after they were on the news.

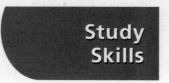
Lesson 5—How Do Changes to Environments Affect Living Things?

Write to Learn

Writing about what you read can help you understand and remember information.

- Many students write about their reading in learning logs. The writing in a learning log can be both creative and personal.
- Writing about the text leads you to think about it.
- Writing your reactions to the text makes it more meaningful to you.

As you read the lesson, pay attention to new and important information. Keep track of the information by completing the learning log below.

Fire Affects Living Things	
What I Learned	**My Response**
Fires can be harmful, destroying animals' habitats.	_____ _____ _____
_____ _____ _____	_____ _____ _____
_____ _____ _____	_____ _____ _____

Name _____

Date _____

Lesson 5—How Do Changes to Environments Affect Living Things?

1. **Investigation Skill Practice–Communicate**

Observe the sign below. Write a sentence to communicate why a forest fire is more likely under these conditions.

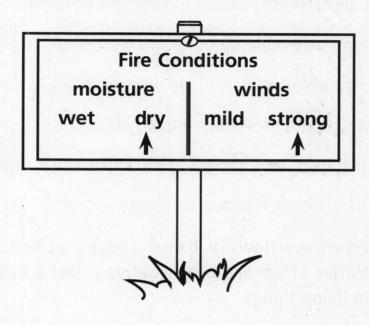

Fire Conditions

moisture   winds

wet   dry   mild   strong

_____

_____

2. **(Focus Skill) Reading Skill Practice–Cause and Effect**

Read the selection. Describe the cause and effects of drought.

It had not rained for more than two months. Rivers and lakes began to dry up. People were not allowed to water their lawns. Plants withered and died.

_____

_____

Use with Unit 3.   (page 1 of 2)   **Science Content Support**   **CS 81**

© Harcourt

Name _____

3. **Check (✓) the statements below that agree with the information found in the lesson.**

    _____ Any kind of plant can grow in the desert.

    _____ Desert wildflowers never live long enough to make seeds.

    _____ Animals need adaptations to survive a drought.

    _____ Fires can help open up more space for plants to get sunlight.

    _____ Ashes from burnt plants help new plants grow.

    _____ Grasses do not grow back quickly after a fire.

    _____ Cutting down trees can affect the balance of a forest.

    _____ The desert has a warm, wet climate.

4. **Sometimes a river overflows its banks, causing a flood. Name one positive effect and one negative effect a flood could have on living things.**

    _____

    _____

    _____

Name _____

Date _____

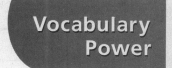

Lesson 6—How Have Living Things Changed over Time?

A. Explore Word Meanings

Write the word from the list with its correct definition.

Ice Age	fossil
petrified wood	extinct

1. _____ a type of fossil where wood turns to stone

2. _____ describes a kind of plant or animal that is no longer living

3. _____ the hardened remains of a plant or animal that lived long ago

4. _____ a time when huge sheets of ice covered parts of Earth

B. Write Sentences

Write a sentence for each of the words above. Use their definitions to help you.

_____

_____

_____

_____

_____

© Harcourt

Name _____

Date _____

Lesson 6—How Have Living Things Changed over Time?

Use a K-W-L Chart

A K-W-L chart can help you focus on what you already know about a topic and what you want to learn about it.

- Use the *K* column to list what you already *know* about fossils.
- Use the *W* column to list what you *want* to learn about fossils.
- Use the *L* column to list what you have *learned* about fossils.

Complete the K-W-L chart as you read this lesson.

Fossils		
What I <u>K</u>now	**What I <u>W</u>ant to Learn**	**What I <u>L</u>earned**
• I know that leaf prints in mud form plant fossils.	• What happens when wood turns to stone? • Why are animal fossils more common than plant fossils?	
• _____ _____ _____ • _____ _____ _____ • _____ _____	• _____ _____ _____ • _____ _____ _____ • _____ _____	• _____ _____ _____ • _____ _____ _____ • _____ _____

Lesson 6—How Have Living Things Changed over Time?

1. Investigation Skill Practice–Use Numbers

Suppose that you are a scientist. You find five fossil footprints of an animal that lived many years ago. You measure the length of each print and record it in the table below.

Print 1	Print 2	Print 3	Print 4	Print 5
10 inches	12 inches	9 inches	4 inches	11 inches

What might you conclude about Print 4? Why?

_____

_____

2. (Focus Skill) Reading Skill Practice–Compare and Contrast

Read the selection. Compare and contrast the two different types of fossils.

Trilobites lived in the sea millions of years ago. They left detailed casts in the rocks where they died. There are no trilobites left on Earth today. Ancient camels also left behind fossils where they died. The camels look like those on Earth today, but were different sizes. Some were about the size of a rabbit, while others were 15 feet tall at the shoulder.

_____

_____

_____

© Harcourt

Name _____

3. Each statement below has a mistake. Rewrite the sentence correctly.

Many cars have been found in the La Brea tar pits.

_____

The woolly mammoth had smooth skin like an elephant.

_____

Fossils show us that no animals on Earth have ever changed.

_____

Footprint fossils look like the animals that made them.

_____

Petrified wood is a living fossil.

_____

Animal fossils are not as common as plant fossils.

_____

The gingko tree has greatly changed in 100 million years.

_____

© Harcourt

Lesson 1—What Planets Orbit the Sun?

Words with Multiple Meanings

Orbit has two meanings. Read the sentences below. Circle the letter that shows which way *orbit* has been used.

1. A moon <u>orbits</u> a planet.

   A. to travel around something

   B. the path an object takes as it travels around something

2. Pluto's <u>orbit</u> is like a stretched out circle.

   A. to travel around something

   B. the path an object takes as it travels around something

Written Cloze Exercise

Study the words and their definitions. Then complete each sentence with one of the words. Use all of the words once.

planet: a large body of rock or gas that orbits a star in space

moon: a large body that orbits a planet

solar system: the sun and the objects that orbit it, including the planets and their moons

telescope: a tool that makes faraway objects seem closer

3. The sun is the biggest object in the _____.

4. If you look through a _____, you can see the rings of Saturn.

5. Earth is the third _____ from the sun.

6. Earth has only one _____.

Name _____

Date _____

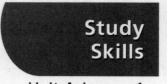

Lesson 1—What Planets Orbit the Sun?

Preview and Question

Identifying main ideas and asking questions about them can help you find important information.

- To preview a lesson, read the lesson title and the section titles. Look at the pictures, and read their captions. Try to get an idea of the main topic, and think of questions you have about the topic.
- Read to find the answers to your questions. Then recite, or say, the answers aloud. Finally, review what you have read.

As you read this lesson, make a chart like the one shown below. Practice reading, reciting, and reviewing.

What Planets Orbit the Sun?				
Preview	**Questions**	**Read**	**Recite**	**Review**
Nine planets, including Earth, orbit the sun in our solar system	Where is Earth located in our solar system?	✔	✔	✔
_____ _____ _____ _____	_____ _____ _____ _____			

© Harcourt

Name _____

Date _____

Lesson 1—What Planets Orbit the Sun?

1. | **Investigation Skill Practice–Use Numbers**

The correct order of the planets, from closest to the sun to farthest from the sun, is Mercury, Venus, Earth, Mars, Jupiter, Saturn, Uranus, Neptune, and Pluto. In the table, write the name of each planet next to its distance from the sun.

Planet	Distance from the Sun
	227.9 million kilometers
	1,427.0 million kilometers
	108.2 million kilometers
	149.6 million kilometers
	2,871.0 million kilometers
	57.9 million kilometers
	4,497.1 million kilometers
	5,913.0 million kilometers
	778.3 million kilometers

2. (Focus Skill) | **Reading Skill Practice–Compare and Contrast**

Read the selection. Compare and contrast Mercury and Venus.

Mercury and Venus are the closest planets to the sun. They are alike in many ways. Both planets are very hot. They have rocky surfaces. Neither planet has a moon. They are also different. Venus is covered by thick clouds. Mercury has no clouds. Venus is more than twice as big as Mercury. Venus takes more than twice as long as Mercury does to orbit the sun.

_____

_____

_____

_____

© Harcourt

Name _____

3. Check (✔) the statements below that agree with the information found in the lesson.

_____ Earth orbits around the sun, which is a star.

_____ In all, thirteen planets orbit the sun.

_____ Earth has several moons that orbit it.

_____ Planets and their moons are part of the solar system.

_____ The sun is at the center of the solar system.

_____ You need a telescope to see the other planets in the sky.

_____ The inner planets are Mercury, Venus, Earth, and Mars.

_____ All of the inner planets are made mostly of frozen gases.

_____ The outer planets are the five planets farthest from the sun.

_____ All of the outer planets have rocky surfaces.

4. Fill in the graphic organizer below. It will help you compare and contrast the inner and outer planets.

	Inner Planets	Outer Planets
Mostly Made of ...	rock	**A**
Number of Moons	**B**	most have many
Distance from the Sun	near	**C**
Size	**D**	most are large

© Harcourt

Name _____

Date _____

Lesson 1—What Planets Orbit the Sun?

The Solar System

A. Fill in the drawing of the solar system. Use the word list in
the box below to fill in the blanks.

Earth	Mars	Neptune	Saturn	Venus
Jupiter	Mercury	Pluto	Uranus	

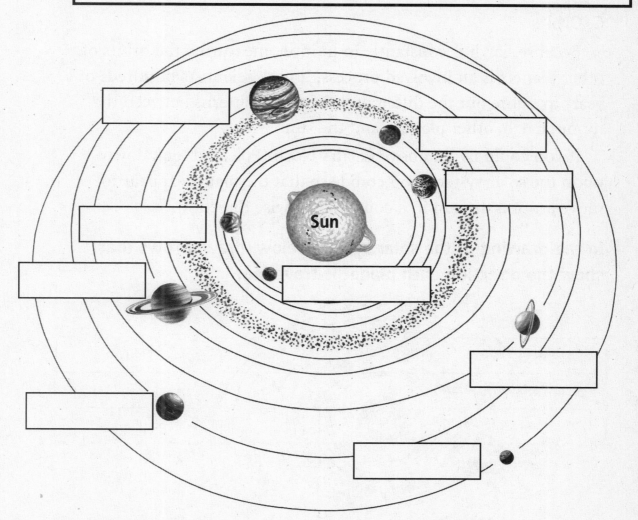

Which of these planets can be seen without a telescope?

_____

Name _____

B. In the drawing below, show the path Earth takes as it orbits the sun. Then show the path the moon takes as it orbits Earth.

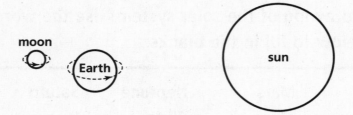

C. Orbits

Because Earth is constantly in motion, measuring the orbits of other planets is an involved process. It took scientists hundreds of years to figure out the different movement patterns between the moon, Earth, other planets, and the sun.

If you could fly far above Earth's North Pole and look down upon the solar system, you could see that the planets appear to move around the sun in a counterclockwise direction.

In the drawing of the solar system below, draw arrows that show the direction each planet is traveling.

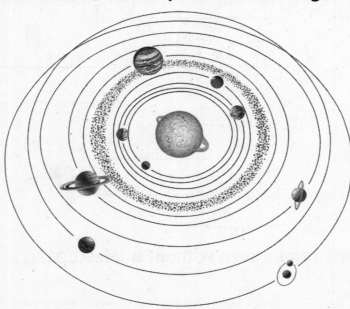

Lesson 2—What Patterns Do Earth and the Sun Follow?

Context Clues

Read each sentence. Think about the meaning of the underlined word. Write the meaning on the line.

1. Earth spins around its <u>axis</u>, similar to the spinning of a toy top or the spinning of a car's wheels around an axle.

_____

2. One <u>rotation</u> of Earth takes 24 hours—or one day. Earth's rotation is similar to the motion of a merry-go-round.

_____

3. Summer is my favorite <u>season</u> because the weather is warm and the days are longer.

_____

4. Earth's axis is <u>tilted</u>, so one half points toward the sun and the other half points away from the sun.

_____

Greek and Latin Roots

Look at the root words and their meanings in the box below. Then write a definition for the words using the roots as a clue.

Hemi is Greek for "half."	*Sphaera* is Latin for "round ball."

5. Northern Hemisphere _____

6. Southern Hemisphere _____

© Harcourt

Name _____

Date _____

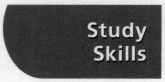

Lesson 2—What Patterns Do Earth and the Sun Follow?

Take Notes

Taking notes can help you remember important ideas.

- Write only important facts and ideas. Use your own words. You do not have to write in complete sentences.
- One way to organize notes is in a chart. Write down the main ideas in one column and facts in another.

Fill in the chart below as you read this lesson.

What Patterns Do Earth and the Sun Follow?	
Main Ideas	**Facts**
Day and Night • The sun lights half of Earth at one time. • Earth makes one rotation in 24 hours.	• It is daytime for parts of Earth that face the sun. • _____ _____
The Sun's Positions • _____ _____ • _____ _____	• _____ _____ • _____ _____
The Seasons • _____ _____ • _____ _____	• _____ _____ • _____ _____

© Harcourt

Name _____

Date _____

Lesson 2—What Patterns Do Earth and the Sun Follow?

1. **Investigation Skill Practice–Draw Conclusions**

Look at the pictures of the sun and the shadows. What conclusion can you draw about how shadows change throughout the day?

morning

noon

evening

_____

_____

_____

2. (Focus Skill) **Reading Skill Practice–Cause and Effect**

Read the paragraph and answer the questions below.

Earth rotates, or spins, once every 24 hours. The spinning of Earth causes the sun to appear to move across the sky. Earth's axis is tilted. When the northern half of Earth is tilted toward the sun, it has summer. The northern half of Earth has winter 6 months later, because then it is tilted away from the sun.

A. What causes the sun to seem to move across the sky?

_____

B. What is the effect of Earth's axis being tilted?

_____

© Harcourt

Name _____

Science Concepts

3. Write *true* or *false* next to each statement.

_____ It takes Earth about $365\frac{1}{4}$ days to orbit the sun.

_____ The sun always shines on half of Earth while the other half is in darkness.

_____ It is the same time everywhere on Earth.

_____ Every evening, the sun sets in the west.

_____ The sun travels across the sky.

_____ Shadows are long in the morning because the sun is low in the sky.

_____ At noon, shadows are short because the sun is overhead.

_____ During winter in California, the Northern Hemisphere is tilted toward the sun.

_____ During summer, the days are longer.

_____ Earth's axis is straight up and down in relation to the sun.

4. How do you know when it is morning for people on the other side of Earth? Explain.

_____

_____

_____

_____

Name _____

Date _____

Lesson 3—What Patterns Do Stars Follow?

Use Definitions

Study the words and their definitions. Then complete each sentence with one of the words. Use all of the words once.

star: a hot ball of glowing gases that gives off energy

constellation: a group of stars that appears to form the shape of an animal, a person, or an object

magnify: to make an object appear larger

galaxy: a large group of stars

Big Dipper: a constellation of seven stars that looks like a bowl with a handle

lens: a curved piece of glass used to bend light and magnify objects

1. Scientists can use a telescope to _____ the rings of Saturn.

2. You can recognize the star pattern called the _____ even though its location in the sky changes.

3. People gave each _____ a name that described the shape of an animal, person, or object.

4. The telescope with the largest _____ can magnify objects the most.

5. The sun is a _____ that provides heat and light to Earth.

6. Earth is part of the Milky Way, a _____ made of millions of stars.

© Harcourt

Name _____

Date _____

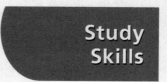

Lesson 3—What Patterns Do Stars Follow?

Use an Anticipation Guide

An anticipation guide can help you anticipate, or predict, what you will learn as you read.

- Use what you know about the subject of each section to predict the answers to the questions in the table below.
- Read to find out whether your predictions were correct.
- You can make a chart and write your own questions on a separate sheet of paper.

Stars Appear to Move		
Question	**Prediction**	**Correct?**
What causes the stars to appear to move?	_____ _____ _____ _____	
Star Patterns and the Seasons		
Question	**Prediction**	**Correct?**
What causes the stars to change position from season to season?	_____ _____ _____ _____	
Telescopes Help Us Observe Stars		
Question	**Prediction**	**Correct?**
Why do scientists use telescopes to observe stars?	_____ _____ _____ _____	

© Harcourt

Name _____

Date _____

Lesson 3—What Patterns Do Stars Follow?

1. **Investigation Skill Practice–Compare**

Suppose two friends use two different telescopes. Jennifer borrows a toy telescope. Akashi borrows a scientist's telescope. Compare what both friends can see when they look at the night sky.

_____

_____

_____

2. **Focus Skill** **Reading Skill Practice–Main Idea and Details**

Read the selection. Underline the main idea. List at least two details about the main idea.

In this lesson, you learned about groups of stars. Groups of stars can be classified in different ways. A small group of stars can be known as a constellation. This is a pattern, made by stars, that forms an imaginary picture in the sky. A galaxy is another kind of group of stars. A galaxy is huge. It can have a spiral, round, or egglike shape, or have no specific shape at all.

_____

_____

_____

© Harcourt

Name _____

3. In the chart below, write the cause or the effect in the blank space.

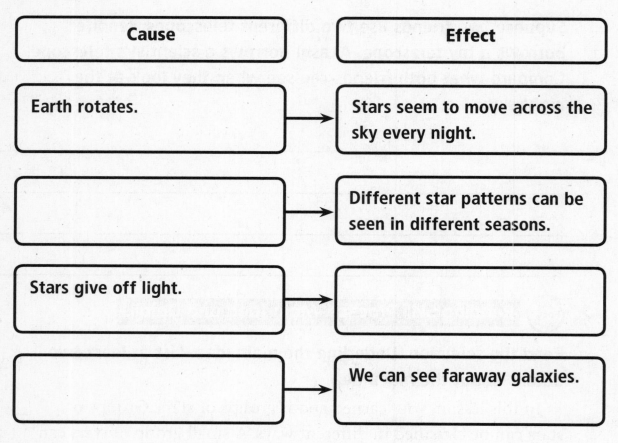

Cause	Effect
Earth rotates.	Stars seem to move across the sky every night.
	Different star patterns can be seen in different seasons.
Stars give off light.	
	We can see faraway galaxies.

4. A sailor is in the middle of the ocean at night. He has no compass to help him get home. How do you think constellations can help the sailor? Explain.

_____

_____

_____

© Harcourt

Name _____

Date _____

Lesson 4—Why Does the Shape of the Moon Seem to Change?

Matching Exercise

Match the clue on the left to the term on the right.

1. _____ when just the edge of the lit part of the moon can be seen

2. _____ the moon phase in which the lighted half of the moon is not visible from Earth

3. _____ when the moon looks like a half circle and is lighted on one side

4. _____ when the lighted part of the moon seems to be getting smaller

5. _____ the different shapes the moon seems to have

6. _____ when the lighted part of the moon seems to be getting bigger

7. _____ the moon phase in which the moon looks like a circle

A new moon
B full moon
C moon phases
D waxing
E waning
F quarter moon
G crescent moon

© Harcourt

Name _____

Date _____

Lesson 4—Why Does the Shape of the Moon Seem to Change?

Connect Ideas

You can use a web organizer to show how different ideas and information are related.

- List important themes in the web's center.
- Add bubbles showing main ideas that support each theme.
- Add bubbles for the details that support each idea.

Copy and complete the web below as you read this chapter.

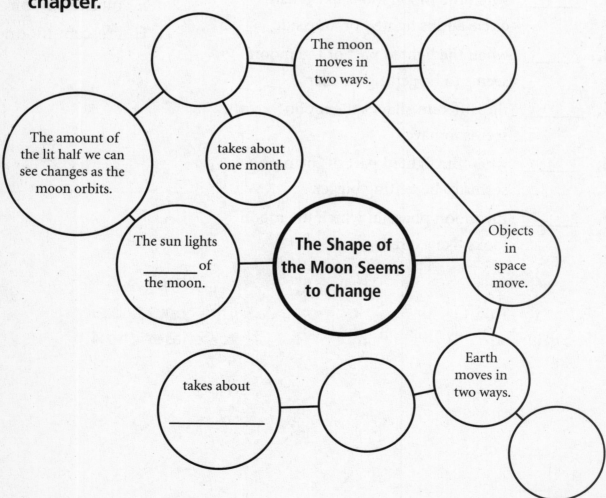

Use with Unit 4.

Name _____

Date _____

Lesson 4—Why Does the Shape of the Moon Seem to Change?

1. **Investigation Skill Practice–Infer**

   Look at the picture of the model below. Make inferences about the model and what it represents.

   _____

   _____

   _____

2. ⭐(Focus Skill) **Reading Skill Practice–Sequence**

   Put the following phases of the moon in the correct sequence, beginning with the new moon. Number the phases 2 to 4.

   _____ third-quarter moon

   _____ first-quarter moon

   ___1___ new moon

   _____ full moon

Name _____

3. Each statement below has a mistake. Rewrite the sentence correctly.

Earth orbits the moon.

_____

It takes the moon one year to travel once around Earth.

_____

The moon makes its own light.

_____

One-third of the moon is lit by the sun.

_____

When the moon is waxing, the part you can see becomes smaller.

_____

During the new moon, you can see half of the moon in the sky.

_____

A crescent moon occurs just before and just after a full moon.

_____

4. Linn went outside and saw a full moon in the sky. About a week later she went outside. Describe and name the type of moon Linn would see. Explain.

_____

_____

_____

© Harcourt

Name _____

Date _____

Lesson 4—Why Does the Shape of the Moon Seem to Change?

Moon Phase Calendar

A. Fill in the calendar with phases of the moon. Imagine that you observed a full moon on the first day of the month. Use the box below to draw and label all the phases of the moon on the calendar. Then color days with yellow when the moon is waxing and color days with blue when the moon is waning.

full moon	new moon	first-quarter moon
third-quarter moon	crescent moon (2)	

APRIL						
Sun	Mon	Tues	Wed	Thurs	Fri	Sat
1	2	3	4	5	6	7
8	9	10	11	12	13	14
15	16	17	18	19	20	21
22	23	24	25	26	27	28
29	30					

After you complete the calendar, answer the following questions.

1. Why is there usually only one full moon each month?

_____

_____

2. How is the crescent moon on April 12 different from the crescent moon on April 18?

_____

© Harcourt

B. In the drawing below, the moon is in different positions as it orbits Earth. Color the sun yellow. Then shade in, using gray, the half of the moon that is dark. Next, shade the half of Earth that is in nighttime. For each of the four drawings, in the right-hand-column draw how the moon appears, from Earth, in the night sky.

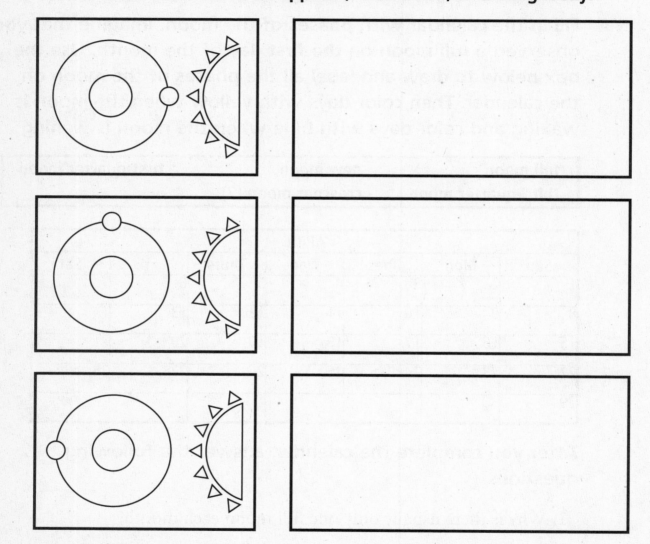

VOCABULARY GAMES
and CARDS

Contents

Vocabulary Games

You can use the vocabulary cards on pages CS111–CS176 to play these games. The cards are provided for each chapter in your science textbook. Each card has a word on one side and the word's definition on the back. For some of these games, you may need to keep the definition hidden from view.

Guess the Word

You will need vocabulary cards, paper and pencil

Grouping large group or pairs

1. Form two teams. Each player must have a partner. One player in each pair is the clue giver and one is the guesser.
2. Teams take turns playing. The first clue giver draws a word card and gives the guesser one clue at a time. A scorekeeper records the number of clues given to guess a word.
3. After the word is guessed, play is passed to the other team. After all cards have been used, the team using the fewest clues wins!

Word Ladder

You will need vocabulary cards, tabletop

Grouping groups or partners

1. Place the cards in a pile, hiding the definitions.
2. Player One chooses a card, reads the word, says the word's meaning, and then turns the card over to check the meaning.
3. If the word is correct, the card is kept and Player One continues until a word is missed. The card is returned to the pile and the next player plays.
4. The player with the most cards is the winner!

You will need vocabulary cards, paper and pencil   Hidden Words

Grouping whole class or large group; small group; pairs

1. Choose a word from the vocabulary cards to hide in a sentence. For example, the word *heat* is hidden in the following sentence: Mit<u>ch</u> <u>eat</u>s ice cream.

2. Write a sentence with a hidden word. Exchange papers with a classmate to find each other's hidden words.

You will need vocabulary cards, grid paper, and pencil   Crossed Words

Grouping small groups or pairs

1. Place the cards face-up on a table so that each word can be seen. Choose one word for your crossword puzzle. Write that word vertically on the grid. Identify it by writing a 1 in the box with the first letter, just like a crossword puzzle. Use the word's definition on the back of the card to help write a clue for this starter word.

2. Choose a second word to add to the grid. Be sure it shares a letter with the first word written. Attach it to the first word by writing the second word horizontally on the grid. Identify it by writing a 2 in the box with the first letter of that word.

3. Continue to attach words to the puzzle and number each word. Write a clue for each of the numbered words.

4. Give a partner a blank grid with spaces numbered to match your puzzle. Help your partner by shading each square that contains a letter. Have your partner solve the puzzle.

You will need 2 identical sets of word cards,
paper and pencil

Grouping small group of at least five

1. One player is named as the "host." The others form two teams. Each team has a set of word cards face-down in the same order.

2. The host asks one person from each team to draw a card, checking to make sure both are looking at the same word. He or she gives a one-word clue about the word. Team members then try to guess the word.

3. The host gives a point if the word is guessed. If the word is not guessed, the player from the other team who saw the word gives a second one-word clue about the word. If a word is not guessed after five rounds, no team gets a point and the host reveals the word to both teams.

4. The team with the most points wins the game.

You will need vocabulary cards,
science textbook

Red Light, Green Light

Grouping large or small groups

1. Each player has a set of the same word cards. One person is the host.

2. The host holds up a word for everyone. Each player then pulls that word card from his or her pile. Players open their books to the chapter the word came from.

3. The host calls out, "green light." Each player quickly scans the text to find the word used in a sentence. When the word is found, that word card is placed in the book to mark the page and the book is closed. That player calls out, "red light." Play then stops.

4. The sentence with the word is read aloud. The player scores one point. Play then continues with the next word.

© Harcourt

absorb

accurate

adaptation

analyze

accurate

[AK•yuh•ruht]

Correct.

Mark measured the root twice to make sure his measurement was *accurate*.

absorb

[ab•SAWRB]

To take in.

The color black *absorbs* all of the colors of light.

analyze

[AN•uh•lyz]

To figure out what the data shows.

A model can help you *analyze* information.

adaptation

[ad•uhp•TAY•shuhn]

A trait that helps a living thing survive.

A giraffe's long neck is an *adaptation* for reaching leaves high on a tree.

atom

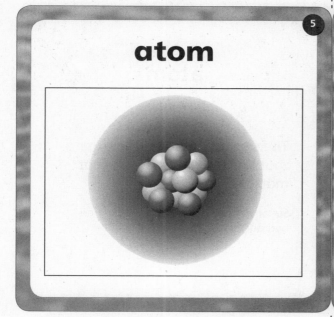

axis

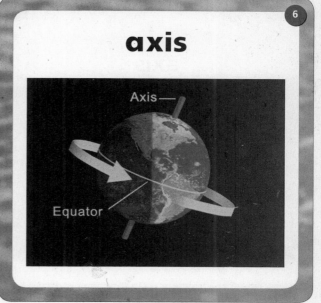

balance

bar graph

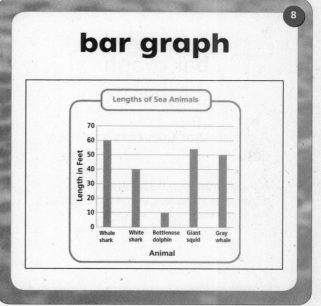

© Harcourt

axis

[AK•sis]

An imaginary line that runs through Earth from the North Pole to the South Pole.

Earth rotates around its *axis*.

⑥

atom

[AT•uhm]

The smallest particle of matter that has the properties of that matter.

Sugar is made of carbon, oxygen, and hydrogen *atoms*.

⑤

bar graph

[BAR GRAF]

A graph that uses bars to display data.

This *bar graph* can help you see differences in the lengths of animals that live in the sea.

⑧

balance

[BAL•uhns]

Not too many and not too few of a kind of living thing.

When people change the environment, they may change the *balance* among living things.

⑦

battery

9

binoculars

10

biome

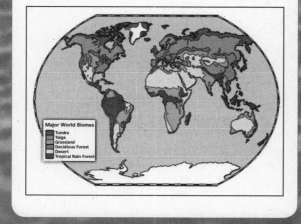

Major World Biomes
Tundra
Taiga
Grassland
Decidious Forest
Desert
Tropical Rain Forest

11

boiling

12

© Harcourt

binoculars

[bih•NAHK•yuh•lerz]

A tool you hold in your hands and look through with both eyes to make distant objects seem closer.

Binoculars work the way a telescope does, but you use both eyes to look through them.

10

battery

[BAT•er•ee]

An object that changes stored chemical energy into electrical energy.

A *battery* contains stored energy.

9

boiling

[BOYL•ing]

A change from a liquid to a gas as a result of being heated quickly.

After *boiling* for five minutes, the pot of water was almost empty.

12

biome

[BY•ohm]

A part of the world with a certain climate and certain plants and animals that can survive there.

This map shows the major *biomes* of the world.

11

© Harcourt

chemical change

13

circuit

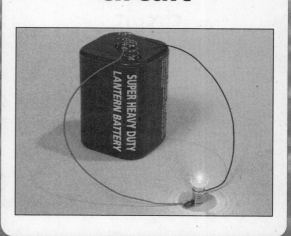

14

climate

15

compare

16

circuit

[SER•kuht]

A path that electricity follows.

Electricity flows through this *circuit*, which connects a light bulb to a battery.

14

chemical change

[KEM•ih•kuhl CHAYNJ]

A change in which two or more substances combine to form a new substance.

Rusting is one kind of *chemical change*.

13

compare

[kuhm•PAIR]

To tell how two things are similar and different.

You *compare* two animals by telling how they are the same and how they are different.

16

climate

[KLY•muht]

The weather a place has over a long period of time.

The *climate* in some parts of California is very cold in the winter.

15

© Harcourt

conclusion

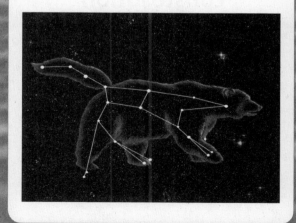

17

condense

18

constellation

19

crescent moon

20

condense

[kuhn•DENS]

To change from a gas to a liquid.

Water in the air condenses when it loses heat and becomes cool.

18

conclusion

[kuhn•KLOO•zhuhn]

A statement based on data that summarizes what was learned in an inquiry.

A conclusion should be written based on what you learn in an experiment.

17

crescent moon

[KRES•uhnt MOON]

The phase of the moon in which just the edge of the lit part can be seen.

A crescent moon is seen just before and just after a new moon.

20

constellation

[kahn•stuh•LAY•shuhn]

A group of stars that appears to form the shape of an animal, a person, or an object.

Ursa Major is a constellation that many people think looks like a bear.

19

© Harcourt

cycle

	Sun	Mon	Tue	Wed	Thr	Fri	Sat
							1.
2.	3.	4.	5.	6.	7.	8.
9.	10.	11.	12.	13.	14.	15.
16.	17.	18.	19.	20.	21.	22.
23.	24.	25.	26.	27.	28.	29.
30.	31.					

data

data table

Height of Plant

	Monday	Wednesday	Friday
Week 1	2 cm	3 cm	5 cm
Week 2	8 cm	9 cm	11 cm

desert

data

[DAY•tuh]

Information.

Data is collected by using measuring tools and recording the measurements.

cycle

[SY•kuhl]

A sequence of events that happens over and over again.

The *cycle* of the moon's phases takes $29\frac{1}{2}$ days.

desert

[DEZ•ert]

A very dry environment.

A *desert* gets very little rain.

data table

[DAY•tuh TAY•buhl]

A display that organizes data in rows and columns.

A *data table* helps you organize your data as you record it.

dissolve

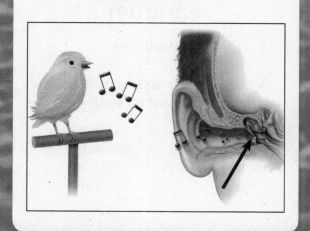

drought

eardrum

Earth

drought

[DROWT]

A long period of time with very little rain.

A *drought* can be bad for farmers because plants need water to grow.

dissolve

[dih•ZAHLV]

To become completely mixed with a liquid.

Sugar *dissolves* in water.

Earth

[ERTH]

The third planet from the sun in the solar system.

We live on the planet *Earth*.

eardrum

[IR•druhm]

A thin, skin-like sheet inside your ear; it vibrates when sound energy strikes it.

Sound waves cause the *eardrum* to vibrate, allowing us to hear sounds.

electricity

element

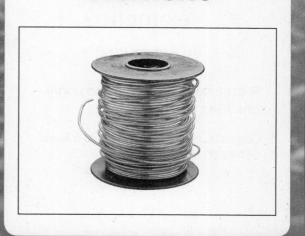

energy

environment

© Harcourt

element

[EL•uh•muhnt]

A pure substance that has only one kind of atom.

Copper is an *element* because it is made of only one kind of atom.

30

electricity

[ee•lek•TRIS•uh•tee]

A kind of energy that moves through a wire.

Electricity runs many of the things we use at home and at school.

29

environment

[en•VY•ruhn•muhnt]

Everything that is around a living thing.

Many kinds of living things can share the same *environment*.

32

energy

[EN•er•jee]

The ability to cause change.

Energy is needed to push a basket across the floor.

31

evaporation

33

evidence

34

experiment

35

extinct

36

evidence

[EV•uh•duhns]

Information collected in a scientific inquiry.

You must collect *evidence* to prove whether or not your prediction is correct.

34

evaporation

[ee•vap•uh•RAY•shun]

The change of state from a liquid to a gas.

The footprints are drying up due to *evaporation* of the water by the sun.

33

extinct

[ek•STINGT]

Describes a kind of thing that is no longer found on Earth.

Dinosaurs are *extinct*.

36

experiment

[ek•SPEHR•uh•muhnt]

A test done to find out if a hypothesis is correct.

In this *experiment*, the only difference between the cups is the type of soil the plants are growing in.

35

first-quarter moon

37

food

38

forest

39

fossil

40

food

[FOOD]

Substances from which plants and animals get energy and nutrients.

Food gives you energy to run and play.

first-quarter moon

[FERST KWAWRT•er MOON]

The moon phase in which the moon looks like a half-circle and is lighted on the right side.

The *first-quarter moon* is one-quarter of the way through its phases.

fossil

[FAHS•uhl]

The hardened remains of a plant or an animal that lived long ago.

We can learn about living things of long ago by studying *fossils*.

forest

[FAWR•uhst]

An area that is covered with trees.

Many kinds of animals live in a *forest*.

freezing 41

friction 42

fuel 43

full moon 44

friction

[FRIK•shuhn]

A force that slows the motion of objects that are touching.

Rubbing your hands together causes *friction*, which warms your hand.

42

freezing

[FREEZ•ing]

The change from a liquid to a solid.

Freezing turns liquid water into ice cubes.

41

full moon

[FUHL MOON]

The moon phase in which the moon looks like a circle.

A *full moon* is very bright in the night sky.

44

fuel

[FYOOL]

A substance that is burned to release stored energy.

Some machines use gasoline as a *fuel*.

43

galaxy 45

gas 46

grassland 47

growth 48

gas

[GAS]

Matter that has no definite shape or volume.

As gas is pumped into the ball, the ball grows larger.

46

galaxy

[GAL•uhk•see]

A large group of stars.

Earth is part of the Milky Way Galaxy.

45

growth

[GROHTH]

Getting larger in size.

You can use a ruler to measure the growth of a living thing.

48

grassland

[GRAS•land]

A dry, flat area on which mostly grasses grow.

A grassland has many animals that eat grass.

47

habitat

49

heat

50

hibernate

51

horizon

52

© Harcourt

heat

[HEET]

Energy that warms objects.

Heat from a stove cooks food.

habitat

[HAB•ih•tat]

The place where something lives in an environment.

In some grassland environments, prairie dogs live in an underground *habitat*.

horizon

[huh•RY•zuhn]

The line where the sky seems to meet the land.

Every evening, the sun sets on the *horizon*.

hibernate

[HY•ber•nayt]

To spend the winter in a kind of deep sleep.

When an animal *hibernates*, its breathing and heartbeat rate slow down and its body temperature drops.

inquiry

investigation

iris

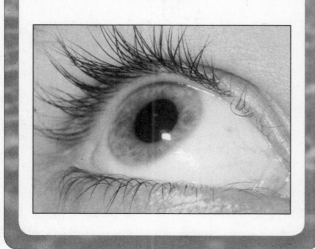

Jupiter

© Harcourt

investigation

[in•ves•tuh•GAY•shuhn]

A scientific study.

The students performed an *investigation* to find out what substances dissolve in water.

inquiry

[IN•kwer•ee]

A question about something or a close study of it.

The students learned a lot from their *inquiry* about greenhouses.

Jupiter

[JOO•pit•er]

The fifth planet from the sun.

Jupiter is the largest planet in the solar system.

iris

[EYE•ris]

The colored part of the eye that controls how much light enters the eye.

The *iris* of this eye is blue.

lens

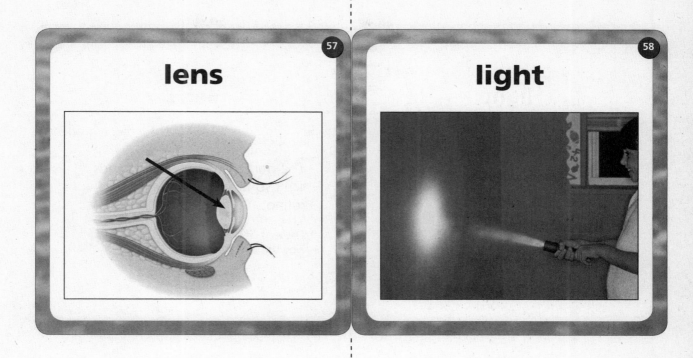

57

light

58

line graph

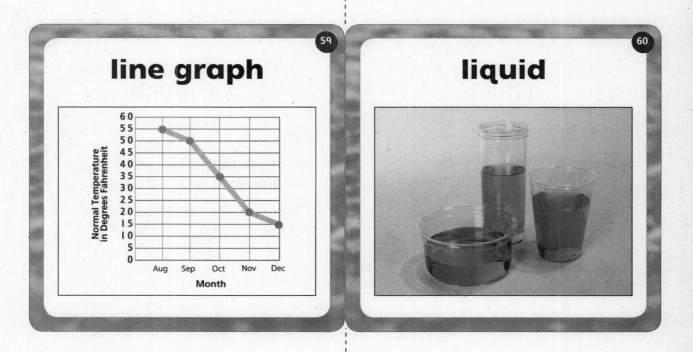

59

liquid

60

© Harcourt

light

[LYT]

A form of energy you can see.

Your eyes can sense light energy.

lens

[LENZ]

The part of the eye that bends light so that it shines on the retina.

The lens of your eye focuses light, the way the lens of a camera does.

liquid

[LIK•wid]

Matter that has a definite volume but no definite shape.

Water is a liquid; its shape can change but its volume does not.

line graph

[LYN GRAF]

A graph that uses lines to display data.

You can see how temperature and time of year are related in this line graph.

logical ⑥¹

machine ⑥²

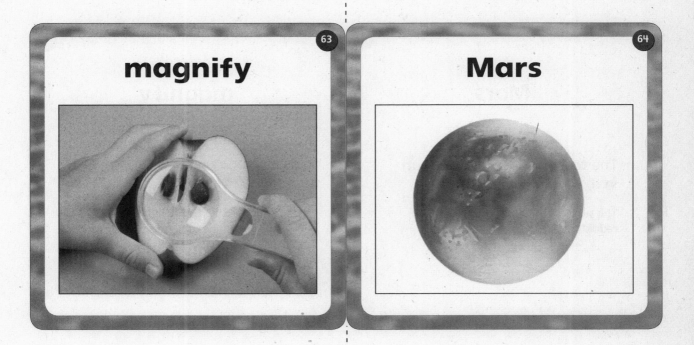

magnify ⑥³

Mars ⑥⁴

machine

[muh•SHEEN]

An invention that uses energy to do work.

Some *machines*, such as car engines, burn gasoline to get energy to do work.

62

logical

[LAHJ•ih•kuhl]

Based on facts; reasonable.

The conclusion was *logical* because it made sense.

61

Mars

[MARZ]

The fourth planet from the sun in the solar system.

The surface of *Mars* is covered with reddish dust.

64

magnify

[MAG•nuh•fy]

To make an object appear larger.

Microscopes, hand lenses, telescopes, and binoculars all *magnify* objects.

63

mass

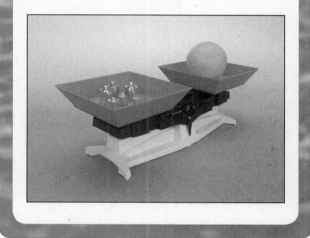

65

matter

66

melting

67

Mercury

68

matter

[MAT•er]

Anything that takes up space.

Everything that takes up space is *matter*, even if the matter is not visible to the eyes.

mass

[MAS]

The amount of matter in an object.

A balance is used to measure how much *mass* an object has.

Mercury

[MER•kyoor•ee]

The planet closest to the sun in the solar system.

The surface of *Mercury* gets hotter than the surface of Earth.

melting

[MELT•ing]

The change of state from a solid to a liquid.

Melting causes water to drip from these icicles.

© Harcourt

migrate

69

mirror

70

mix

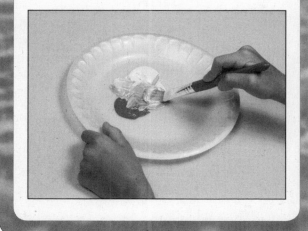

71

mixture

72

mirror

[MIR•er]

A polished surface that reflects light.

You can see your reflection in a *mirror*.

70

migrate

[MY•grayt]

To travel from one place to another and back again.

Some kinds of birds *migrate* south for the winter and return north for the summer.

69

mixture

[MIKS•cher]

A combination of two or more kinds of matter.

Lemonade is a *mixture* of water, sugar, and lemon juice.

72

mix

[MIKS]

To combine things.

You can *mix* colors of paint to make new colors.

71

© Harcourt

moon

moon phases

Neptune

new moon

moon phases

[MOON FAYZ•uhz]

The different shapes the moon seems to have.

The *moon phases* change as the moon orbits Earth.

moon

[MOON]

A large body that orbits a planet.

The *moon* orbits Earth about every 4 weeks.

new moon

[NOO MOON]

The moon phase in which the lighted half of the moon cannot be seen from Earth.

It is easier to see stars at night when there is a *new moon*.

Neptune

[NEP•toon]

The eighth planet from the sun in the solar system.

Since *Neptune* is the next-to-last planet, it has very cold temperatures.

Northern Hemisphere

77

ocean

78

opaque

79

opinion

80

Cats are the best pets.

ocean

[OH•shuhn]

A large body of salty water.

Whales live in the ocean.

78

Northern Hemisphere

[NAWR•thern HEM•ih•sfir]

The northern half of Earth.

California is in the Northern Hemisphere.

77

opinion

[uh•PIN•yuhn]

A personal belief that is not based on evidence.

Make sure you don't confuse your opinions *with evidence that comes from an investigation.*

80

opaque

[oh•PAYK]

Describes an object that does not let light pass through it.

You cannot see through opaque *objects.*

79

orbit

81

particle

82

periodic table

83

planet

84

particle

[PART•ih•kuhl]

A very small bit of something.

All matter is made of tiny *particles* called atoms.

orbit

[AWR•bit]

To travel around an object.

It takes about $365\frac{1}{4}$ days for Earth to *orbit* the sun.

planet

[PLAN•it]

A large body of rock or gas that orbits a star in space.

Saturn is one of nine *planets* that orbit the sun.

periodic table

[pir•ee•AHD•ik TAY•buhl]

A chart that shows the elements.

Elements are arranged in order in the *periodic table*.

Pluto

pollution

predict

property

85

86

87

88

pollution

[puh•LOO•shuhn]

Harmful material that is added to the environment.

Oil spills are a kind of *pollution* that can kill many living things.

Pluto

[PLOOT•oh]

The ninth planet from the sun in the solar system.

One year on *Pluto* is as long as 248 years on Earth.

property

[PRAHP•er•tee]

A characteristic of matter that you can observe or measure.

One *property* of lemons is their sour taste.

predict

[pree•DIKT]

To tell what you think will happen in the future.

You *predict* the outcome of an experiment before you begin the experiment.

pupil

89

react

90

reflect

91

release

92

react

[ree•AKT]

To combine and form a new substance.

When these yellow and clear liquids *react*, *they form a new red substance.*

pupil

[PYOO•puhl]

The opening in the front of the eye, through which light passes.

The pupil *of your eye looks like a black dot.*

90

89

release

[rih•LEES]

To give off.

Charcoal releases *heat when it is burned.*

reflect

[rih•FLEKT]

To bounce off a surface.

Light reflects *off the shiny surface of a mirror.*

92

91

reproduce

retina

rotation

Saturn

© Harcourt

retina

[RET•uh•nuh]

The back of the inside of the eye, where images are formed.

A nerve carries information from the *retina* to the brain.

reproduce

[ree•pruh•DOOS]

To produce new living things.

Living things *reproduce* in many ways, such as by making seeds, laying eggs, or giving birth to live young.

Saturn

[SAT•ern]

The sixth planet from the sun in the solar system.

Saturn has a system of rings spinning around it.

rotation

[roh•TAY•shuhn]

The spinning of Earth.

One *rotation* of Earth takes 24 hours.

scientific method

97

season

98

seismograph

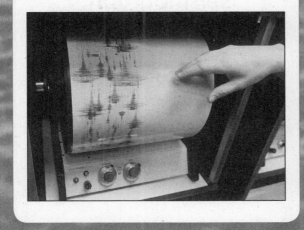

99

shadow

100

season

[SEE•zuhn]

A time of year that has a certain kind of weather.

Spring, summer, fall, and winter are the four *seasons*.

98

scientific method

[sy•uhn•TIF•ik METH•uhd]

An organized plan that scientists use to conduct an investigation.

This student is using the *scientific method* to design an experiment.

97

shadow

[SHA•doh]

A dark area that forms where an object blocks the path of light.

When you move, so does your *shadow*.

100

seismograph

[SYZ•muh•graf]

A tool that records the strength of energy waves in an earthquake.

A *seismograph* can detect earthquakes that are not strong enough for us to feel.

99

solar system

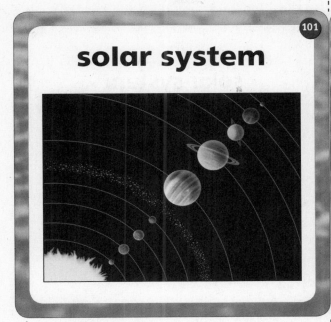

101

solid

102

sound

103

Southern Hemisphere

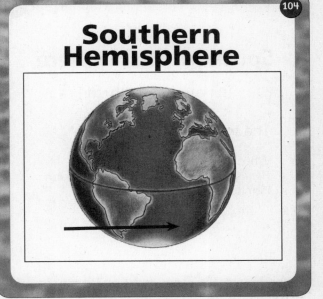

104

solid

[SAHL•id]

Matter with a definite volume and shape.

Marbles, pasta, and grains of sand are *solids*, so their shapes and masses do not change.

solar system

[SOH•ler SIS•tuhm]

The sun and the objects that orbit it, including the planets and their moons.

Nine planets orbit the sun in the *solar system*.

Southern Hemisphere

[SUHTH•ern HEM•ih•sfir]

The southern half of Earth.

When it is winter in the *Southern Hemisphere*, it is summer in the Northern Hemisphere.

sound

[SOWND]

Energy that can be heard.

The *sound* made by a drum is different from the sound made by a trumpet.

© Harcourt

species

star

state

stationary

star

[STAR]

A huge ball of hot, glowing gases that gives off energy.

There are many *stars* in the night sky.

106

species

[SPEE•sheez]

One kind of living thing.

Wildebeests are one *species* living in Africa.

105

stationary

[STAY•shuh•nair•ee]

Not moving; fixed in one location.

Earth is not *stationary* but is constantly rotating on its axis as it orbits the sun.

108

state

[STAYT]

One form that matter can have.

Solid, liquid, and gas are three *states* of matter.

107

© Harcourt

summer

109

sun

110

survive

111

symbol

112
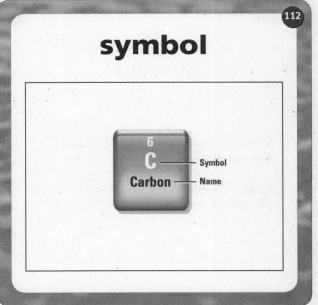

© Harcourt

sun

[SUHN]

The star at the center of the solar system.

The *sun* provides light and heat to Earth.

110

summer

[SUHM•er]

The season with warm weather and the longest days.

You can wear lighter clothing during the *summer* months.

109

symbol

[SIM•buhl]

One or more letters that stand for an element.

C is the *symbol* for carbon on the periodic table of the elements.

112

survive

[ser•VYV]

To stay alive.

Plants and animals have adaptations that help them *survive*.

111

© Harcourt

telescope

113

third-quarter moon

114

tilted

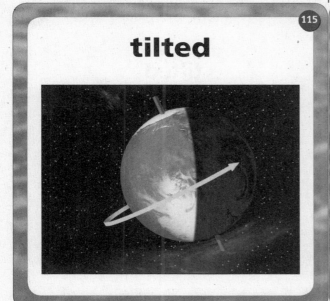

115

trait

116

third-quarter moon

[THERD KWAWRT•er MOON]

The moon phase in which the moon looks like a half-circle and is lighted on the left side.

The third-quarter moon is three-fourths of the way through its phases

114

telescope

[TEL•uh•skohp]

A tool that makes faraway objects seem closer and larger.

A telescope helps you see more clearly objects that are far away.

113

trait

[TRAYT]

A characteristic, or feature, of a plant or animal.

The different traits of these flowers help you tell them apart.

116

tilted

[TILT•uhd]

Leaning in one direction; not straight up and down.

Earth's axis is slightly tilted.

115

© Harcourt

transfer

translucent

transparent

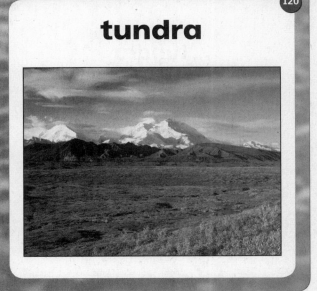

tundra

© Harcourt

translucent

[tranz•LOO•suhnt]

Describes an object that lets
some light pass through it.

You can see light through *translucent*
objects, but you cannot see clearly
through them.

118

transfer

[TRANZ•fer]

To carry energy or cause it to
move from one thing to another.

The bowling ball *transfers* energy to
the pins.

117

tundra

[TUHN•druh]

A place that has a very cold
climate.

Animals that live in the *tundra* all year
have adaptations to survive in the cold.

120

transparent

[tranz•PAIR•uhnt]

Describes an object that lets
most light pass through it.

You can see clearly through *transparent*
objects.

119

universe

Uranus

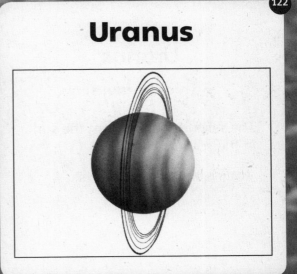

Venus

vibrations

Uranus

[YOOR•uh•nuhs]

The seventh planet from the sun in the solar system.

Uranus is the third-largest planet.

122

universe

[YOO•nuh•vers]

Everything that exists, including stars, planets, and energy.

Many galaxies make up the *universe*.

121

vibrations

[vy•BRAY•shuhnz]

A series of back-and-forth movements.

The *vibrations* of a bell cause sound waves to move through the air.

124

Venus

[VEE•nuhs]

The second planet from the sun in the solar system.

You do not need a telescope to see *Venus* in the night sky.

123

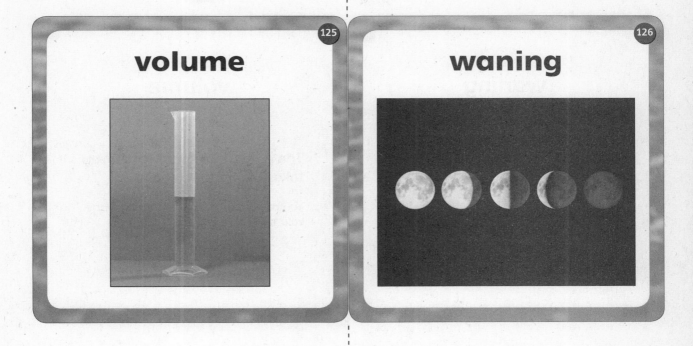

volume 125

waning 126

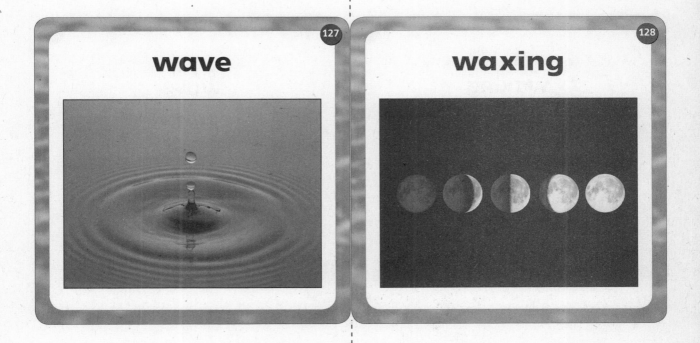

wave 127

waxing 128

© Harcourt

waning

[WAYN•ing]

Becoming smaller.

We say the moon is *waning* during the phases between a full moon and a new moon.

126

volume

[VAHL•yoom]

The amount of space something takes up.

You can use a graduate to measure the *volume* of a liquid.

125

waxing

[WAKS•ing]

Getting bigger.

We say the moon is *waxing* during the phases between a new moon and a full moon.

128

wave

[WAYV]

A disturbance that moves energy to other places.

Light travels in *waves* that are similar to waves in water.

127

wetland

129

winter

130

© Harcourt

winter

[WIN•ter]

The season with cold weather and the shortest days.

Many trees lose their leaves in winter.

wetland

[WET•land]

An area that is often flooded.

Many kinds of ducks live in wetlands.